THE
IRON
MEN

D0228236

THE
IRON
MEN
THE WORKERS WHO
CREATED THE NEW
IRON AGE
ANTHONY BURTON

The
History
Press

Cover illustrations © Shutterstock

First published 2015

The History Press
The Mill, Brimscombe Port
Stroud, Gloucestershire, GL5 2QG
www.thehistorypress.co.uk

British Library Cataloguing in Publication Data.
A catalogue record for this book is available from the British Library.

ISBN 978 0 7509 5955 1

Typesetting and origination by The History Press
Printed in Great Britain

CONTENTS

ACKNOWLEDGEMENTS

The author would like to thank the following for permission to reproduce illustrations (the remainder are from the author's collection):

Elton Collection, Ironbridge Gorge Museum, 41; Mansell Collection, 129; National Museum of Wales, 41; Pamlin Prints, 85; Richard Thomas & Baldwin, 138; Scottish record office, 105; Shropshire Record Office, 64; Waterways Archive, Gloucester, 63 (bottom) and 66; Windfall Films, 91.

1

THE FIRST IRON AGE

Since cast-iron has got all the rage,
And scarce anything's now made without it;
As I live in this cast-iron age,
I mean to say something about it.
There's cast-iron coffins and carts,
There's cast-iron bridges and boats,
Corn-factors with cast-iron hearts,
That I'd hang up in cast-iron coats.

We have cast-iron gates and lamp-posts,
We have cast-iron mortars and mills, too;
And our enemies know to their cost
We have plenty of cast-iron pills, too.
We have cast-iron fenders and grates,
We have cast-iron pokers and tongs, sir;
And we soon shall have cast-iron plates,
And cast-iron small-clothes, ere long, sir.

These are just two verses from a broadsheet ballad published in 1822 with the satirical title *Humphrey Hardfeatures' Description of Cast-iron Inventions.* There are three more verses listing still more cast-iron inventions, some real, others imaginary. The song is quoted not for its very limited literary merits, but because it is an indication of just how important iron had become by the beginning of the nineteenth century, so common a part of everyday life that it was seen as an appropriate subject for a popular song sheet. Cast iron was indeed transforming the world, and not merely cast iron: the other two forms – wrought iron and, to a lesser extent, steel – were also having a profound impact on society as a whole.

The world had entered a New Iron Age, whose influence was arguably to have a far greater impact than the prehistoric period that was first given the name. This book is about the men and women who made this great transformation possible and who worked with the different forms of iron. But to understand exactly what it was that made the new age possible, you have to know just how it differed from the earlier period. We have to step back in time, not just a few years or even centuries, but through the millennia.

There are two intriguing questions to ask about the traditional Iron Age. The first is, how did anyone ever discover that heating a lump of rock could produce a metal – something as unlike stone as it is possible to get? The second is, having done that to produce silver, lead, copper and tin, why did it take nearly another 1,500 years before anyone found a practical way of producing iron that could be used for a whole range of different tools and artefacts?

No one can really answer the first question, other than to say silver objects that date back to around 3000 BC have been excavated in Egypt and Mesopotamia. But if you look at the ore galena, it has a silvery metallic sheen that might suggest it would be worth experimenting with it to see what would happen. We know that it is primarily lead sulphide but associated with a little silver and that by roasting it the sulphur can be burned off as sulphur dioxide, leaving metallic lead and silver behind. Once you have heated one interestingly coloured rock it makes sense to try others to see what they might produce.

Copper was almost certainly originally found by people looking for gold. Near the gold deposits were dark nodules with a greenish tinge, and anyone taking the trouble to scratch away the surface would have found native copper buried in the centre. These small lumps of metal were quite difficult to work, but the process could be improved by annealing – heating to a high temperature and then slowly cooling.

It was not, perhaps, too big a step to discovering that there were very promising ores, notably copper pyrites, which looked as if they might also produce something worthwhile. However, it was not simple to reduce them to metal. There must have been a long period of experimentation, as heating alone is not enough and eventually they would have found that two stages were necessary. First, the ore had to be roasted, and then heated in some sort of furnace, which would need to be provided with a blast of air to raise the temperature to the point where the molten metal could be produced. The technology of smelting had been born.

The copper ores were frequently found associated with tin, and the combination of these two metals produced the alloy bronze. This had all sorts of advantages over pure copper, with greater strength and hardness, making it valuable for a whole range of products from tools and weapons to ornaments. It was such a huge leap in technology that its introduction was used to mark the start of a whole new period of civilization – the Bronze Age, which began around 1600 BC and lasted for over 1,000 years. So, throughout this period metal extraction and working was developing and still iron played, at best, a minor part in the story.

Which brings us to our second question. Why didn't the production of iron get going during all those centuries? The short answer is that it did, but only on a very small scale. Iron could be found naturally in meteorites, but in such small quantities that it was useless for anything much more than ornaments. It took a long time to make the connection between this iron and its ore. Like the other ores, iron ore looks as if it ought to produce something interesting when heated in a furnace, but when it was first tried the result must have been very disappointing. What appeared at the end of the process would have appeared rather like a fossilised sponge, a lump of stone riddled with holes. This is known as a 'bloom' and the actual iron is hidden away inside a mass of slag and cinder.

It takes repeated heatings and hammerings to turn the bloom into wrought iron. Even when iron did appear, it was not immediately obvious what to do with it. Bronze could be given a hard cutting edge by hammering. Do that with cold iron and you do get an edge but it blunts quite easily, much more easily than bronze, so there was no obvious incentive to develop the technology. It was the discovery that a different technique could be used which made the great breakthrough. To produce a hard, sharp edge requires heating, hammering while still red hot and then quenching the hot metal in cold water. The result was a metal that was actually superior to bronze in terms of hardness and durability, and a new age was born.

In Britain, the Iron Age is conventionally described as lasting from the end of the Bronze Age, at around 500 BC, to the arrival of the Romans. The dates simply reflect the fact that archaeologists have given these labels – Stone Age, Bronze Age and Iron Age – specifically to the prehistoric period before written records. In terms of technology, this first Iron Age could be said to have lasted far longer, right up to the beginning of the sixteenth century.

Technology looks very different, with the benefit of hindsight, from the way it appeared to the workers of the past. We know, thanks to modern scientific investigation, that wrought iron is a very pure form of the metal. Seen under the microscope, it has a fibrous structure, which allows it to be bent and shaped with comparative ease. The aim of the early ironmasters was to produce this form of the metal, which they could work and use to make essential items such as tools and weapons, and everyday objects like nails and horseshoes.

The early furnaces, known as 'bloomeries', in which iron was made were comparatively crude. At their simplest they were no more than depressions in the ground, which were then filled with the ore and charcoal for use as a fuel, which would be covered by a dome of clay or some other fireproof material. More sophisticated versions would be constructed like stubby chimneys, also made of some sort of fire-resistant material. The resulting bloom would have been quite small. Because the nature of iron was not understood, there was no way of working out how long the ore should be kept in the bloomery, apart from trial and error. If it was not long enough, the reduction would be incomplete. If it was left in too long, some of the carbon would infiltrate the bloom, and what emerged was not wrought iron, but cast iron.

We know that cast iron contains roughly 4–5 per cent of carbon which, in the bloomery, would have come from the charcoal during the overlong heating process. We also know that under the microscope it looks very different from wrought iron. Now the structure is crystalline. It is brittle, and attempts to bend and shape it prove futile: it simply snaps. This would have been more than a little irritating to the iron makers who now had a form of iron for which they could find very little use. The search began to find a better system that would ensure they got just what they needed – wrought iron, and nothing else. It took a very long time to come up with the answer. The blast furnace was probably introduced in the region around Liège sometime around 1400, but only reached Britain a century later.

The shape of the first blast furnaces were similar to two truncated pyramids, one stuck on top of the other, with the smaller, inverted pyramid being at the bottom. Starting at the base was the hearth, a flat bed of stone, above which the furnace widened out to an area known as the 'boshes', before narrowing in again. It would be open at the top to allow fuel and ore to be added. A small opening in the hearth area allowed a pipe (the 'tuyere') to be inserted, through which air could be blown to raise the temperature of the furnace. In the small bloomeries this could be done by using hand-operated bellows, in exactly the same way as a blacksmith would increase the temperature of the hearth in his forge.

But the blast furnaces were on a much larger scale, so the bellows were too big to work by hand and had to be powered by a waterwheel. During the smelting process, impurities would also appear. It was found that by adding limestone to the charge, a liquid slag could be produced. As this was lighter than the molten metal, it could be tapped off separately. In these early years, no one had any use for the slag, so it was simply discarded, and the smouldering slagheap became a familiar part of the industrial landscape.

At the back of the furnace was an arch, across which the material for the furnace could be wheeled and tipped into the open top. The inside of the furnace was lined with fire-resistant material. Originally it was square in cross-section, but in later furnaces it tended to be circular. This can be clearly seen in the surviving remains of a seventeenth-century blast furnace – the Coed Ithel furnace on the wooded hillside above the Wye, north of Tintern.

Once the furnace was alight it could, in theory, be kept going for years, apart from stoppages made from time to time for repairs to the lining. A continuous blast of air would be passed through the tuyere, and charges of charcoal, ore and limestone added at regular intervals as the molten metal and slag were tapped off.

This was only the start of the process. The iron was run off into troughs. There was a central gully, the runner, from which subsidiary troughs ran, and a further set of small channels ran off these. The latter were thought to look like sows feeding their piglets, hence the name 'pig iron'. This was the unwanted cast iron, which was now taken to the 'finery' to be converted into wrought iron. The finery was not unlike the familiar blacksmith's hearth, with the charcoal fire heated to a high temperature by means of an air blast. The cast iron was added to the fire, then a second blast of air was blown over the hearth. The oxygen in this second air blast combined with the carbon in the cast iron to be carried away as carbon monoxide, leaving the pure metal behind.

The blast furnace had several advantages over the bloomery. Because the slag was removed in the smelting process, there were fewer impurities and greater quantities of metal could be produced. This created the need for new ways of working the iron. With large quantities, the blacksmith's strong right arm was no longer enough, so water-powered hammers were introduced. The early hammers were all tilt hammers. The actual hammer was pivoted at its centre, and the tail of the hammer came into contact with projections on a rotating drum, turned by the waterwheel. Each time a projection hit the end of the hammer, it pushed it down, lifting the hammer head. As the projection was cleared, the hammer head fell back onto the metal on the anvil.

An iron-working site had become a complex affair involving many stages, each of which required a waterwheel, to pump air or to activate hammers. So sites needed to have a guaranteed regular supply of water. Streams were dammed to create ponds, many of which still exist long after their use has largely been forgotten.

The most important of all the areas was the Sussex Weald, and if you look at an Ordnance Survey map of the area you will find certain names cropping up all over the region: 'Hammer Pond', 'Furnace Pond' and 'New Pond'. The Weald had another attraction for ironmasters. It was densely wooded. Furnaces consumed vast amounts of charcoal: estimates for one ironworks measured its requirements in the somewhat vague units of wagon loads, and that came out at 1,800 loads a year. This was just the amount of charcoal needed, and producing that required an even greater quantity of timber. An efficient charcoal maker would be able to produce 1 ton of charcoal from 4 tons of timber so, in effect, the ironworks consumed over 7,000 loads of timber every year. That represented a huge area of woodland to be felled. This did not necessarily mean woodland was being destroyed. Coppicing was used, in which the wood supply was regenerated on a regular basis. Even so, it placed a huge strain on the country's woodlands, and the ironmasters were not the only ones chopping down trees in vast quantities.

One of the greatest commentators on life in Britain in the early years of the eighteenth century was Daniel Defoe, best known these days as the author of the novels *Robinson Crusoe* and *Moll Flanders.* But in his lifetime, he was even better known as a political commentator and recorder of the social scene. In his book *A Tour through the Whole Island of Great Britain*, he described the country as he travelled through it in the years 1724–1726. He visited the Weald and wrote:

> I had the curiosity to see the great foundaries [*sic*], or iron-works, which are in this country, and where they are carry'd on at such a prodigious expence [*sic*] of wood, that even in a country almost all over-run with timber, they begin to complain of the consuming of it for these furnaces, and leaving the next age to want timber for building their navies.

Defoe rather pooh-poohed the notion that the supply of timber might run out, but it was a fact that shipbuilders, not just the navy, also used prodigious amounts of wood. When a wooden man-of-war was built the amount of timber was again measured in wagonloads, and a typical big ship was estimated to need 2,800 loads. One vessel, the *Great Michael*, built in 1801, was said to have 'wasted all the woods in Fife that were oak woods'. It was a serious problem, and limited the expansion of the iron industry.

New technologies for developing different uses for the iron were slow to evolve. The most important invention was the slitting mill which, like the blast furnace, probably first came into use in the area round Liège in the early sixteenth century. The history of technology is littered with tales of industrial espionage and the story of this invention is no exception. Fable has it that news of the slitting mill reached England, but just how it worked was a closely guarded secret. A Midlands ironmaster, Richard Foley, disguised himself as a wandering musician, visited Liège and took secret notes, came home and set up in business in 1628. It is a good story and it seems a shame to spoil it, but sadly it is pure invention.

One known fact about the introduction of the technology to Britain is that, far from the citizens of Liège keeping the process secret, it was one of their own citizens, Godfrey Box, who set up a mill at Dartford in 1595, long before the Midland troubadour went on his supposed wanderings. There may even have been slitting mills at an earlier date in the Wye Valley. It seems odd these days, when the area round Tintern Abbey is thought of as simply a romantic spot of great natural beauty, to think that at the beginning of the seventeenth century it was one of the busiest iron-making and iron-working sites in the country.

Prince Pückler-Muskau wrote an account of his travels in Britain, titled *Tour in England, Ireland and France in the Years 1828, 1829, 1830*, when the Tintern ironworks were still very active. 'Fires gleam in red, blue, and yellow flames, and blaze up through lofty chimneys, where they assume the form of glowing flowers.' He was impressed by the immense waterwheel. 'The frightening noise when it was first set going, the furnaces around vomiting fire, the red hot iron, and the half-naked black figures brandishing hammers and other ponderous instruments, and throwing around the red hissing masses, formed an admirable representation of Vulcan's smithy.' The prince's description might be a bit short of technical details, but it gives a wonderful picture of what must have been a very dramatic scene.

The original mills and fineries produced large bars of metal, but there was a demand for smaller bars for making all sorts of essential items, especially nails. The first part of the process consisted of passing the iron bar through a pair of water-powered rollers to flatten it. After that it was passed through rotary cutters to form it into strips. A similar process was used to make wire, which involved attaching an iron rod to a crank, turned by a waterwheel. This was used to draw the rod through holes of diminishing size in an iron plate. One of the major uses was in making needles: one end of the wire was bent over to form a loop and the other sharpened to a point.

Cast iron was still the poor relation in the iron family. Even when objects might have seemed obvious candidates for its use it was largely ignored. Large guns were constructed from wrought iron. Bars of iron were arranged around a central core, welded together and then strengthened by shrinking iron hoops round them – much as a cooper made a barrel (although the technology has changed, the name 'gun barrel' is still used). But apart from specialist uses, a great deal of iron simply went to blacksmiths who could make a wide range of objects in their own smithies.

But what of the third member of this family – steel? This is a form of iron with a carbon content somewhere between that of cast iron and the pure metal. It was extremely difficult to make, but highly valued for its hardness and ability to be sharpened to a fine edge without losing its strength. The finest steel was imported and generally known as 'Damascus' steel, and swords made from this material were greatly prized. The name was given because the swords came from the Middle East, but the actual steel was made even further east, in India. This steel, later to be known as 'wootz' steel, was manufactured in clay crucibles filled with iron, together with specially chosen green twigs, covered by leaves and then sealed with clay. The crucibles were then heated in a furnace. The resulting steel was of exceptional quality and recent analysis has suggested that it might be due to nanoparticles of carbon and carbides within the metal. Its manufacture was a craft passed down through generations, but making steel in crucibles did not reach Britain until centuries after the secret was discovered in the East. We tend to think that advances in technology are the prerogative of western nations. Wootz steel is a reminder that, in some cases, we lagged far, far behind the technologists in the east.

This, then, was the situation in Britain by the seventeenth century. It has been estimated that there were less than 100 charcoal blast furnaces at work, and roughly four times as many forges. Wrought iron was the most common product and the one most in demand.

Technology does not stand still, and all the time more uses were being found for iron and demand was increasing. The big problem was how to provide the extra material without stripping the country of its woods and forests. Attempts were made by Dud Dudley, the illegitimate son of Lord Dudley, who had extensive works in the Black Country, to use coal for smelting. He claimed some success in a book, *Metallum Martis*, which was published in 1665, but in general, coal introduced far too many impurities to make usable iron in any shape or form. Dudley took out a patent for using coal, but it contained no useful information to give a hint of how he achieved the results he had claimed. Whether he had been successful or not, no one else was able to make coal smelting work. The real breakthrough came half a century later, and it was this development that marked the start of the New Iron Age.

2

THE DARBYS OF COALBROOKDALE

Coalbrookdale, in Shropshire, has been called 'the birthplace of the Industrial Revolution' and awarded World Heritage Site status, and it is not difficult to see why. It was here that two great technological breakthroughs occurred that made the Industrial Revolution possible.

This new world could not have come into existence without a guaranteed supply of iron being available to manufacture the machinery on which it was based. One has only to consider the rapid spread of new textile machinery introduced in the eighteenth century to appreciate that this one industry alone would be requiring vast amounts of this essential raw material. And, as the Industrial Revolution progressed, machines of ever greater complexity were being invented.

It was no longer just a case of making iron available, but the regular supply also had to be accompanied by a technology that would enable that complexity to be accurately reproduced. The best way of ensuring that the same parts could be turned out over and over again was by making an accurate pattern from which identical castings could be reproduced. Cast iron was finally coming into its own, thanks to the inventiveness of the ironmasters of Coalbrookdale.

However, the narrative doesn't start in Shropshire and the early part is confused by a certain amount of mythology and technical inaccuracies being mixed into the earliest accounts of exactly what happened.

The story begins with Abraham Darby, who was born in Dudley in the heart of the Black Country in 1678. His father was a nail maker and a Quaker, and he was apprenticed to another Quaker, Jonathan Freeth of Birmingham, who had a business making malt mills for the brewing industry. Darby's religious beliefs were to have an important part in both his business and private life.

At the end of his apprenticeship in 1699 he married Mary Sargeant, the daughter of a linen yarn bleacher, and shortly afterwards they moved to Bristol, where Darby set up in business on his own, still making malt mills. Over the years, the Darbys were to have ten children, only four of whom – Abraham, Edmund, Ann and Mary – survived into adulthood.

In 1702 Darby went into partnership with three other Quakers to establish a brass works at Baptist Mills, Bristol. Throughout his career, his partnerships and collaborations tended to be made among the Friends. This was a rapidly developing industry in Bristol, based on the fact that all the important ingredients were available in the West Country – zinc from the Mendips, copper from Devon and Cornwall and coal for the furnaces from the Somerset Coalfield.

The brass was shaped into a variety of cooking pots by hammering. At some time, probably in 1704, he seems to have visited Holland, where brass pots were being cast in sand, and he brought back a number of Dutch workers to Bristol. By this time he had also set up an iron foundry at Cheese Lane in Bristol. Now a certain amount of confusion sets in, as some of the accounts seem to get muddled as to which of the works – Baptist Mills or Cheese Lane – were the source of Darby's experiments with casting pots.

An account was written by Hannah Rose, the daughter of John Thomas, who was to play a key role in the story. According to this version, Darby tried to get his Dutch employees to apply their casting techniques to the iron industry. What he was hoping to do was manufacture bellied iron pots, the sort of cauldrons associated with the popular images of witches – Macbeth's weird sisters, for example, with their chant of 'fire burn and cauldron bubble' – and cartoon cannibals boiling missionaries.

According to Hannah Rose, the Dutch were unsuccessful. It was her father, John Thomas, who as a young man had come to Darby to learn the trade of malt mill making and asked to be allowed to try and develop the technique. His efforts greatly impressed Darby, who realised that they were working on a potentially lucrative process and that it was essential to keep this to themselves until it could be patented. Hannah Rose wrote:

> His Master and him were bound in Articles in the year 1707 that John Thomas should be bound to work at that business and keep it a secret and not teach anybody else, for three years. They were so private as to stop the keyhole of the door.

The agreement allowed Thomas a wage of £7 per annum and £8 for the last year of the agreement, together with 'good and sufficient meat, drink, washing and lodging'. If Hannah Rose is correct in giving her father most of the credit for developing the process, then it seems a meagre reward, bearing in mind that Thomas was already 28 years old and that Darby was well aware that this could be a very profitable enterprise, as the wording of the Patent, taken out the same year, makes clear. It stated in the preamble exactly what Abraham Darby had achieved 'by his study and industry and expence':

> He hath found out and brought to perfection a new way of casting iron bellied pots, and other bellied ware in sand only, without loam or clay, by which iron pots, and other ware may be cast fine and with more ease and expedition, and may be afforded cheaper than they can be by the way commonly used, and in regard to their cheapness may be of great advantage to the poore of this our kingdome, who for the most part use such ware, from whence great quantities are imported, and likewise may in time supply foreign markets with that manufacture of our own dominions.

The patent did not go unchallenged, but Darby successfully saw off the opposition, gaining for himself a virtual monopoly in a commodity that was in use in households throughout the land. This was a time long before such items as gas cookers had even been dreamed of, and many families still cooked over an open fire. The pots had three legs so that they could stand on the floor with the fire beneath them. The patent stressed the value of the process for making pots, but gave little hint that this was a process that would prove of immense importance for centuries and with far wider applications.

Casting in sand involves creating a pattern, usually of wood, which is placed in a special casting box, then packed round with sand. Not just any sand can be used: it has to be a variety which, when dried out, can retain the shape of the pattern. *A Manual of Foundry Practice*, published as recently as 1948, lists three particular types of sand then in use, including the greensand that was used by Darby. This might suggest that just one ingredient is used, more or less in a natural state. This is very far from being the case.

Nearly all sands have to be dried to just the right consistency and cleaned of all other material such as clay. Greensand would normally be a mixture of two types of sand, with an addition of 6–10 per cent coal dust. Darby was fortunate that very suitable sands were available at Kidderminster and Stourbridge, close to the navigable River Severn.

The special quality of casting sands mean that they can be used not just for rather crude objects like bellied pots, but also for very intricate items, such as machine parts. Making the patterns for these was a highly skilled job for a carpenter. I have a pattern for a large cogwheel sitting on the window-sill in front of me, which was actually used at the Coalbrookdale Foundry. It is 7in diameter and has thirteen teeth. The central part of the cog is not one solid cylinder, but made of three separate segments and each of the teeth is curved and fitted into the core by a dovetail joint. It has obviously been made with great care. It still has its original works' number 231 carved into it. Somewhere in the foundry there would be a record book setting out just what machine this would have formed a part of. This same pattern could be used time and again to produce the part for new machines, or as a replacement if the original casting became worn or broken.

Darby's partners were not greatly enthusiastic about the idea of branching out into the iron-founding business, so he began looking for somewhere he could set up a new business on his own. He found the ideal site at Coalbrookdale in Shropshire. It had been the works of an ironmaster called Shadrach Fox, and contained an old charcoal blast furnace in a poor state of repair. But it was an excellent situation, close to the River Severn, which was navigable by the big sailing barges known as 'trows', right up to a wharf at the end of Coalbrookdale, just a short distance from the works. (The river was such an important transport route that, at this period, it was known as 'the King's Highway of Severn'.)

There was a good supply of essential raw materials, including extensive woodland for charcoal making and a reliable source of water for power. Although charcoal was available in considerable quantities, the price had been steadily rising for some time. The foundry was also surrounded by coalfields, and it is possible that Darby was already beginning to think of different ways of fuelling his new furnace.

Having started his life in the family home in Dudley and having worked his apprenticeship in nearby Birmingham, Darby would have been aware of Dud Dudley's claims to have used coal for smelting – and would also have been aware that it did not really work. But he also had experience of a different industry – malting. This process involves heating partially germinated barley in a kiln to dry out. The fuel used affects the taste of the barley – the peat fires used in some Scottish distilleries, for example, give the Scotch a very distinctive flavour. The Bristol maltsters would have discovered that using coal in the furnaces certainly did alter the taste – and not in a way that anyone enjoyed, giving it a sulphurous tang. They found that they could get round this by turning the coal into coke, burning off the sulphur as sulphur dioxide.

At this time, coke was made in a process not unlike that for making charcoal. Great heaps of coal were allowed to smoulder rather than burn, driving off all the unwanted gases. It seems a logical step to wonder if, since coke did not taint barley, it might be possible to use it in a blast furnace without contaminating the iron. It was certainly worth the experiment, and Darby knew he could always fall back on the old method of using charcoal if the idea did not work.

The process of coking was described by one of those eighteenth-century tourists, who had a curiosity to look into all things new, and to describe them in detail. The grandly named B. Faujas de Saint-Fond wrote an account of the process as he saw it at a foundry in Scotland in his 1799 book, *Travels in England, Scotland and the Hebrides*:

> A quantity of coal is placed on the ground, in a round heap, of from twelve to fifteen feet in diameter, and about two feet in height. As many as possible of the large pieces are set on end, to form passages for the air; above them are thrown the smaller pieces, and coal dust, and in the midst of this circular heap is left a vacancy of a foot wide, where a few faggots are placed to kindle it. Four or five apertures of this kind are formed round the ring, particularly on the side exposed to the wind … as the fire spreads, the mass increases in bulk, becomes spongy and light, cakes into one body, and at length loses its bitumen, and emits no more smoke. It then acquires a red, uniform colour, inclining a little to white; in which state it begins to break into gaps and chinks, and to assume the appearance of the underside of a mushroom. At this moment the heap must be quickly covered with ashes … to deprive it of air.

Darby's first task was to rebuild the old furnace – and it was genuinely old. The furnace was tapped through an arch topped by metal beams, one of which carried the date 1638. This was retained through the renovation process and subsequent alterations, and can still be seen today. The hearth was later extended and the present structure dates from the final rebuild and is recorded with the inscription 'Abraham Darby 1777'. The name is that of the founder's grandson, Abraham Darby III. Now the centrepiece of the Coalbrookdale Museum of Iron, this is the furnace from which the first iron to be smelted using coke poured in 1709.

Darby was fortunate, or perhaps one should say wise, in his choice of sites. Everything was organised to ensure maximum efficiency. The stream that provided the power was dammed to create a series of furnace ponds terraced into the valley above the works. Coal and iron pits were also above the level of the works, so the raw materials could easily be brought downhill to the furnace. The finished products then had another downhill journey to the Severn.

At first, the products mostly consisted of pots and pans. Samuel Smiles, the great recorder of many aspects of industrial history, took the trouble to go through the original 'Blast Furnace Memorandum Book' of 1713, and found that the works were turning out 5–10 tons of iron a week:

> The principal articles were pots, kettles and other 'hollow ware', direct from the smelting furnace; the rest of the metal was run into pigs. In course of time we find that other castings were turned out: a few grates, iron pestles and mortars, and occasionally a tailor's goose. The trade gradually increased until we find as many as 150 pots and kettles cast in a week.

The business was thriving, but it was not yet feeding anything approaching an industrial revolution, simply supplying goods for a strictly domestic – in all sense of the word – market.

Abraham Darby died in 1717, when his son Abraham Darby II was still only 6 years old, so the management of the works passed to Richard Ford. He was to remain in control right through to 1745, a period which saw some major changes in the works. He developed the old forges alongside the works and, although the main output was still in pots and pans, he greatly extended the range of customers.

The original Coalbrookdale Works had made little impact on the world at large – a provincial ironworks turning out the most humble of objects was never going to arouse huge interest. Now Ford was operating on a much grander scale. Where Darby had largely supplied comparatively local demands, Ford was now sending out wares as far as London and Cornwall, Cumberland and Northumberland.

The original Darby furnace at Coalbrookdale.

His main customer was a Bristol merchant, Nehemiah Champion, who developed an extensive export trade. He was taking as much as 30 tons of pots and pans in every shipment from the works. A typical order for one month in 1730 was for 2,370 pots and kettles and 550 'small ware'. The works were thriving, but now there were new products starting to appear that would help to place Coalbrookdale on the national industrial map. By 1735, young Abraham Darby II was taken into the partnership and was responsible for what was to prove to be an important new element.

By the beginning of the eighteenth century, the steam engine was starting to have an impact on the industrial world. The first really successful version was the work of Thomas Newcomen, and was initially used purely for pumping water from mines. It was a massive and comparatively crude machine. Pump rods were hung from one end of an overhead beam, and a piston fitted into a cylinder was suspended from the opposite end. Steam was passed into the cylinder below the piston and then condensed by spraying with cold water, creating a vacuum. Air pressure now forced the piston down, at the same time raising the pump rods at the opposite end of the beam. Pressure equalised, the weight of the pump rods then dragged that end down again and the whole cycle could be repeated – the beam nodding up and down, pump rods rising and falling.

It was young Darby's idea to install an engine at the works to pump water from the furnace ponds back up the hillside. This got rid of one of the problems plaguing all water-powered sites – drought that could bring the whole place to a halt. Water was no longer allowed to run away to waste. Thanks to the Newcomen engine, mines could now go deeper than ever before, ensuring coal stocks would not run out and coke-fired furnaces would not be troubled by fuel supplies.

But the arrival of the steam engine had an even more important effect on the industry. Initially, the cylinders had been cast out of expensive brass. Coalbrookdale began casting them in iron. It never became a huge part of their output, but iron castings were to be vital to the whole story of the development of steam power, which would culminate in the next century in the development of the locomotives and railways. Coalbrookdale was to play an important part in that story as well.

During Ford's time, the enterprise expanded, with new furnaces being built at Bersham and Willey. When he died in 1745 Abraham Darby II took over full control and he began looking for new markets. There was a constant demand for iron suitable for slitting, especially for use by nail and chain makers. This was easily supplied by using the iron from charcoal furnace fineries, but the pig iron from coke-smelted iron proved far less useful.

Darby experimented constantly, and on one famous occasion he is said to have spent six days and nights at the furnace and when the iron finally proved to be usable he collapsed and had to be carried home. His wife, Abiah Darby, gave an account of what happened next. Darby sent some of his pig iron, together with some charcoal-smelted iron, to the slitting mills without any indication that one batch was any different from the other. When no one spotted the difference the experiment proved to be a success, and a new furnace was added to the stock. She also described the importance of the innovation and her husband's attitude to his new discovery. Writing in 1775, she was describing events that had taken place twenty-six years earlier:

> Edward Knight Esq. A capital Iron Master urged my Husband to get a patent, that he might reap the benefit for years of this happy discovery: but he said he would not deprive the public of Such an Acquisition which he was Satisfyed it would be, and so it has proved, for it soon spread and many Furnaces both in this Neighbourhood and Several other places have been erected for this purpose.

> Had not these discoveries been made the Iron trade of our own produce would have dwindled away, for woods for charcoal became very scarce and landed Gentlemen rose the price of cord wood exceeding high – indeed it would not have been to be got. But from pit coal being introduced in its stead the demand for wood charcoal is much lessen'd, and in a few years I apprehend will set the use of that article aside.

It was not quite the breakthrough that Abiah Darby suggested: there were still problems to be overcome before the pig iron from coke-fired furnaces could successfully be transformed into usable wrought iron on a large scale.

This was a period of steady expansion, and a new figure appears in the story. Richard Reynolds represented one of the Coalbrookdale investors, Thomas Goldney, and he moved into the area in 1756. The following year he married Hannah Darby and took a one-third share in the ironworks at Ketley, just 3 miles from Coalbrookdale. He was to become an innovative ironmaster in his own right, as well as an important figure in helping to prevent the parent company from collapsing into bankruptcy during the trade slump caused by the Seven Years War that began in 1756. He will appear again later in the story.

The works were not only expanding their output, they were also extending the range of products that they made. One very significant item added to the list of castings was iron rails. The idea of running trucks along specially made tracks had been around for a long time. It was a great improvement over moving things along the roads of the period, which were often riddled with potholes and, in winter, were a morass of cloying mud. These tracks, known as tramways, were particularly prevalent in the coal mining districts, especially in the north-east of England.

At first, they were simply made of wood, but these were soon suffering from wear and tear. They could be improved by laying strips of iron along the top, but ultimately the answer was to make iron rails. Experiments in the eighteenth century suggested that a cart on an ordinary road could carry slightly over half a ton, but on iron rails this rose dramatically to 8 tons.

The early rails were very different from modern track. The trucks had plain wheels, as on other carts and wagons, and not the flanged wheels of today's rolling stock. The rails themselves were L-shaped in cross-section, the wheels running on the flat iron and kept in place by the vertical section. With the spread of canals in the late eighteenth century, more and more tramways were being constructed to bring goods overland to the waterways. Coalbrookdale had its own tramway system, and it was here that a world first may, or may not, have taken place in 1803.

The Cornish engineer Richard Trevithick had constructed a steam locomotive for use on the roads in 1801 and had produced a more sophisticated version, looking like a bizarre form of steam-powered mail coach that had been given trial runs in London. Then, in August 1802, he wrote a long and involved letter to his friend and adviser on scientific matters, Davies Gilbert, at the end of which, almost as an afterthought, he added this intriguing sentence: 'The Dale Co. have begun a carriage at their own cost for the real-roads [*sic*] and is forceing [*sic*] it with all expedition.' This would have been the first steam locomotive in the world to be specially built to run on a railway. Yet there are no contemporary accounts of its running. Many biographers have suggested that the Coalbrookdale people were only contemplating the idea and never actually gave it a trial.

A drawing exists, however, of what might be the Coalbrookdale engine, dated 1803. William Reynolds was at Coalbrookdale and was a man always keen to advance technology and try out new ideas. His nephew W.A. Reynolds, writing some time after the event, seemed quite certain that the event did take place:

> There was a beautifully-executed wooden model of this locomotive in my Uncle William Reynolds's possession, which was given me by the widow, the late Mrs Reynolds of Severn House after his death. I was then a boy fond of making model engines of my own, and I broke up this priceless relic to convert it to my own base purpose, an act of which I now repent as if it had been a sin.
>
> The Coalbrookdale engine is, I believe, the first locomotive engine on record intended to be used on a railroad. The boiler of it is now to be seen in use as a water tank at the Lloyd's Crawstone Pit and the fire tubes and a few other portions of it are now in the yard at the Madeley Wood Works. I never knew how it came to be disused and broken up.

Another writer, in a pamphlet published in 1884, described finding a cylinder 'cherished as a valuable relic' of the engine, and gave its dimensions as 4¾ inches with a 3ft stroke. This fits in very well with a drawing dated 1803, which could well have been for this engine.

The big puzzle is why this momentous event was not reported at the time. A possible answer can be found in an event that took place many miles away in Greenwich. The boiler of one of Trevithick's high-pressure stationary engines blew up with disastrous consequences, killing three men and injuring a fourth. There was a powerful faction, led by the steam pioneer James Watt, who were totally opposed to any use of high-pressure steam. They claimed that all these engines were inherently dangerous and made a great deal out of the Greenwich affair, even though the accident was down to an act of appalling negligence on the part of a boy supposed to be overseeing the engine. If they could make a huge fuss about an engine rooted to the spot, what would they think of one designed to chug off on rails all over the countryside?

There is also another possibility: it was not a success. In the first public trial held the following year in South Wales, an engine built in 1804 worked as planned, but broke the brittle cast-iron rails. If the same thing happened at Coalbrookdale, Reynolds might simply have decided to cut his losses and move on to another project. In the event, Trevithick never had any success trying to persuade the world that steam locomotives were the transport system of the future, and it was not until 1814 that the world's first successful railway opened for business. Whether a locomotive did run at the works or not, there was enough evidence to convince the Ironbridge Gorge Museum to build a replica based on the 1803 design. It steams regularly up and down a short length of track at the Blists Hill museum site.

We have run a little ahead of the chronology, and it is time to turn back to the 1770s, when Abraham Darby III was in control. He was the man behind the project that made Coalbrookdale world-famous, the construction of an iron bridge across the Severn Gorge (now the Ironbridge Gorge).

Plans were laid in 1773 for the construction of a bridge, and the necessary approach roads were approved by Parliament. Various ideas were put forward, but the Act gave no details on how the bridge should be built. After quite a lengthy discussion period, the matter was left in the hands of Abraham Darby III. He came up with the radical idea of constructing a single, soaring, high-arched bridge to be built entirely from cast iron.

What is so extraordinary about the proposal is that no one had ever tried to build an iron bridge on anything like this scale, if indeed anyone had built one at all. Yet this was no tentative attempt at learning how to build in iron, but a huge engineering challenge. It was to have a span of 100ft and to rise 40ft above the river. The result, as we all know, is the majestic bridge that still stands today and was so famous that it gave its name to the town that grew up alongside it, Ironbridge.

A detail of the famous Iron Bridge across the Severn, showing how members slot through each other, held by iron wedges.

Up to this time, there were really only two types of materials used for bridges: stone or brick, built up to make a solid arch, or timber. There was obviously no point in trying to emulate a stone construction, using lots of cast-iron blocks, so Darby based his design on the other available model, the timber bridge. Wooden bridges are held together using a variety of joints, mainly dovetail or mortice and tenon, and that is exactly the technique Darby chose to use. He cast sections of the bridge so that they fitted together just as if they were wood, not iron, and in some cases they were designed so that one member could pass through another.

The final structure consisted of interlocking sections, with joints held together and tightened using either screws or wedges. Some of the curved ribs were so large (80ft long) that the old furnace had to be adapted by enlarging the hearth and the moulding floor extended. The new lintel was inscribed with the date 1777 to mark the start of casting. The structural details can mostly still be clearly seen. What is not immediately obvious is that the deck above the ribs was constructed out of iron plates. This was originally covered by a mixture of iron slag and clay to form the roadway.

It became a great tourist attraction, and visitors flocked to see it. One of these, John Byng, was enthralled: 'But of the iron bridge over the Severn, which we cross'd and where we stopped for half an hour what shall I say? That it must be the admiration, as it is one of the wonders of the age.' Those who took their carriages over the bridge paid a toll of 1*s*, but presumably thought it money well spent.

Another visitor, Richard Gough, did more than admire, he gave details about how it was constructed:

> Over the Severn in this Dale was laid 1779, a bridge of cast iron, the whole of which was cast in open sand, and a large scaffolding being previously erected, each part of the rib was elevated to a proper height by strong ropes and chains, and then lowered till the ends meet in the centre. All the principal parts were erected in three months without any accident to the work or workmen or the least obstruction to the navigation of the river.

The bridge was not a one-off wonder. Its construction marked a whole new start in the development of the iron industry. It was a triumphant demonstration that the material could be used in the construction industry. It made use of the special property of cast iron, the fact that it was very strong under compression, and was therefore ideal as a load-bearing material. This use was to be developed in many different ways over the next 100 years, and more.

Coalbrookdale had become famous. Tourists came not only to stare at the now famous bridge, but also to see the dramatic spectacle of furnaces belching flame and pouring out streams of molten metal. Writers described it in terms of the current cult of the picturesque. Arthur Young, who primarily toured Britain to look at the progress of agriculture, took time off from contemplating wheat and turnips to visit the site. He began by describing the natural beauties of the dale itself, with its wooded hillside, and then continued:

> Indeed too beautiful to be much in unison with that variety of horrors art has spread at the bottom: the noise of the forges, mills, &c. with all their vast machinery, the flames bursting from the furnaces with the burning of the coal and the smoak of the lime kilns, are altogether sublime, and would unite well with craggy and bare rocks.

Others were simply horrified by what they saw. The poet Anna Seward wrote a long poem couched in the classical allusions popular at the time, in which she had the home of Naiads 'usurpt by Cyclops'. The beginning sets the tone:

> Scene of superfluous grace, and wasted bloom,
> O, violated Colebrooke [*sic*]! ...

Love it or deplore it, Coalbrookdale was seen as somewhere extraordinary, outside the experience of most of those who came to wonder at it.

For those who worked there it was neither picturesque nor sublime, simply a place where they could earn a living. In some ways, it was much like other industrial enterprises that were burgeoning at the time. But there was one difference; because the owners were Quakers, the relationship between masters and men was subtly changed.

The Darbys and their Quaker partners always gave preference to other Quakers when recruiting their workforce. The Friends' meetings were held in the company offices. It was very different from the established church, where a strict hierarchy was maintained and the gentry had their reserved pews, usually finished in a far grander style than those of the rest of the congregation. The Quakers met as equals; there were no distinctions made at meetings between employers and employees.

Continuity in management was matched by continuity in employment. The wage books show the same names appearing in the ledgers for generation after generation. That does not mean that the Coalbrookdale workers had an easier life than other industrial workers at that time. Reynolds was also a Quaker, but the rules he imposed on the workers were as strict as those set in place by any other industrialist. The following are typical examples:

> Each Person employed in the Works shall come to, and be engaged in the Employment appointed for them, and at their proper Places from the hour of Six in the Morning to Six in the Evening, Breakfast and Dinner excepted, or forfeit the Sum of one Shilling in Addition to the Time lost by Absence or Indolence, which will be deducted from his Weekly Wages.
>
> Those who stay longer at their Meals than the Time allowed, viz: half an Hour at Breakfast and an Hour at Dinner shall forfeit a quarter of a Day's Wages.
>
> If any One is found in that part of the Work where his business does not call him, he shall forfeit One Shilling.

Conditions at Coalbrookdale may have been more stable than in some organisations, but it did not prevent the men there becoming involved in the widespread riots that began in 1756 and spread through the country in protest at the rising cost of provisions. A few of Darby's men took part, but the ringleaders threatened to destroy the works if others did not join the protest. The family offered them money to keep them happy, and also provided provisions for the crowd.

Hannah Darby, in a letter to her aunt, described the violence and plunder that went on in many places, but also placed it in context. The rioters:

> ... called at our house but did not offer any violence – several hundred had meat and drink this time – we baked three days together and sent several miles for it besides, for there was not a bit of bread nor corn nor flower [*sic*] to be had for money, for some miles about – so that the country was in the greatest distress. The mob gave themselves the title of levellers and so they were indeed.

When the uprising had been suppressed, the ringleaders were sentenced to death. Four Coalbrookdale men had been caught up in the riots and were each sentenced to transportation for fourteen years. Darby put in a plea for mercy on their behalf, but the records do not tell if it was successful. From his plea and Hannah's letter it is clear that the Darbys were among the few who appreciated that the rioters had been driven by desperation more than by malice. It might also be significant that Coalbrookdale did not suffer in the way that others did at the hands of the rioters. Perhaps this is another indication that it really was a very special place.

What began with a single furnace at Coalbrookdale had spread out to the surrounding area and had become a great centre of innovation. The enterprise was no longer turning out humble cooking pots, but was now producing anything from machine parts to iron rails. In the nineteenth century it would also develop a reputation for producing ornamental ironwork, culminating in an incredibly ornate fountain specially cast for the Great Exhibition of 1851 at the Crystal Palace.

The area centred on Coalbrookdale would not maintain a monopoly in this new form of iron making for long. The advantages of smelting with coke were so obvious that others were soon following the Darby family's lead. The new ironworks were no longer being built in forested areas, where timber was available for charcoal making. Now they moved to regions close to productive coalfields. The furnaces of the Weald went into decline and new industrial centres were developed in areas such as Scotland and South Wales.

3

THE INDUSTRY DEVELOPS

One of the main obstacles to future progress lay in the fact that, in spite of Darby's efforts, the problem of supplying large quantities of wrought iron from the pig iron produced in the new blast furnaces had not really been solved. Three men from Coalbrookdale made their own attempts and came very close to finding the solution.

The first was made by the brothers George and Thomas Cranage, who patented their idea in 1766, and the second by Peter Onions, who obtained a patent in 1785. Both patents called for the use of a reverberatory furnace, one in which the fuel and the metal being treated never came into direct contact, but neither process ever came into widespread use.

The answer was to appear from the opposite end of the country, and was discovered by a man who had come into the industry more or less by accident. Not a lot is known about Henry Cort's early life, other than that he was born at Lancaster, probably in 1741. The only other individual recorded locally with that slightly unusual name was also called Henry Cort, a mercer and one-time mayor of Kendal, who died in 1747. He was presumably a relation, possibly his father. Our Henry Cort must have moved south at quite an early age, as he was still only 16 when he was working as a clerk for a naval agent in London. He progressed rapidly up the ladder, and by 1761 the naval agency had become a partnership, Batty & Cort. By 1763 he had set up in business on his own and felt he was doing well enough to get married the following year.

Naval agents played an important part in naval affairs, dealing with all kinds of matters, including paying crews, arranging allowances and sorting out the prize money due from the capture of enemy ships. We do not know what happened to his first wife, apart from the fact that she died and that Cort remarried in 1768. His new bride, Elizabeth Haysham, had important family connections, as her father was the steward to the Duke of Portland, a prominent Whig politician.

There are two versions of how Cort came to become involved in an ironworks. In the first version, he had loaned money to a Mr Morgan, who owned the works, and when Morgan was unable to repay the debt, took the works instead. The other, probably more reliable, version appears in the *Dictionary of National Biography*. In this account, the works were owned by his wife's uncle, William Attwick, who supplied ironware to the Royal Naval Dockyard at Portsmouth. Cort invested in the business and eventually, in 1776, took over the management and set about building up its trade.

The main works at Fintley, near Farnham, suffered from a poorly managed water supply, and Cort's first task was to sort that problem out. He was able to use his well-established connections with the navy to secure a valuable contract, mostly for supplying the service with the iron hoops that fitted round the masts of sailing ships. Previously, the ironwork had been forged in the traditional way, using tilt hammers and water power. Cort had different ideas. He developed a system based on rollers. This was not new in itself, but he realised that by cutting grooves in the rollers, he could produce different shapes, not just the flat plates. For example, he could produce rounded rods. He took out a patent for his grooved rollers in 1783.

Cort was working entirely with wrought iron and, like everyone else, found supplies difficult to come by. So he turned his attention to the second problem: how to get wrought iron from the iron pigs produced by the new blast furnace. The basic idea was not very different from that already tried by Onions and the Cranage brothers. He too used a reverberatory furnace, but an altogether improved process.

The problem remained the same: how to burn off the unwanted carbon, while at the same time not introducing any new impurities. Cort's furnace consisted of two iron boxes, both of which were lined with firebrick. The smaller of the two had firebars to take a coal fire. The second box had a chimney at one end, was filled with pig iron and was separated from the other by a firebrick wall. The chimney created an updraught, pulling the flames from the firebox over the cold iron, gradually melting it, while at the same time, drawing air over the molten surface. It was the action of the oxygen in the air coming into contact with the metal that burned off the carbon as carbon dioxide.

The metal had to be kept stirred with metal rods passed through holes in the brickwork, and as the process continued the molten material became stiffer and stiffer, until it began to form into clods. The iron makers called this 'coming into nature', an indication that the process was coming to completion. The stirring, known as 'puddling' was incredibly hard and uncomfortable work. The men had to work close to the intense heat of the flames and, as the metal came to nature, so the job of stirring it became more and more difficult. It may have been hard work, but it was an undoubted success and Cort patented the process in 1784.

Cort realised that it was not enough just to take out a patent and wait for the ironmasters of Britain to come knocking at his door. In the 1780s he toured widely around the country, travelling as far as Scotland, promoting his invention and setting up demonstrations. It seemed that Cort's fortune was made when the Navy Board put out new contracts for bar iron, and announced that they would only consider iron made using the Cort process.

There was no doubt that Cort had used his old naval contacts very effectively, too effectively as it turned out. His partner in the enterprise was Samuel Jellicoe, son of Adam Jellicoe, the deputy paymaster of the navy. It all looked a little too cosy to be altogether comfortable, and investigation suggested that the Cort–Jellicoe enterprise had been partially funded from naval funds. This, if not actually embezzlement, was dangerously close to it. Cort escaped criminal prosecution, but he had to repay £27,500, a huge sum in those days, equivalent to around £1.5 million in today's money. Not surprisingly, Cort was unable to find that amount and in 1789 was officially declared bankrupt. His friends rallied and managed to get him a government annuity in recognition of his work in transforming a vital industry, but it was only a modest £200 a year. He died in lodgings in London in 1800 having made, and lost, a fortune.

As a bankrupt, his patents were forfeited to the state, which took very little interest in them and, as a result, ironmasters around Britain were able to exploit the invention with impunity. The only concession made was a collection among the leading men in the industry to raise a fund to support his widow. Efforts by his sons – he had thirteen children altogether – to persuade the government to help out failed completely. The Cort family never managed to make any money out of their father's invention that had helped to revolutionise a whole industry.

By the end of the eighteenth century, the Industrial Revolution was developing with extraordinary speed. Industry after industry had moved from cottage and home to mill and factory, yet all were still relying on the natural power sources of wind and water. The Newcomen engine had used steam, but had not used the power of steam pressure: all the work was done by the atmosphere pressing down on the piston, and gravity pulling down the pump rods to raise it again. The steam itself was a passive agent: its only role was to be condensed back into water.

That all changed thanks to the work of a Scots instrument maker, with probably the most famous name among all the early inventors and industrialists – James Watt. Popular mythology has it that the young Watt saw the lid of a kettle being lifted and promptly realised the power of steam and invented an engine. Reality is rather more prosaic. He was working as an instrument maker at Glasgow University when he was sent a model Newcomen engine that had failed to perform. This led him to investigate the whole principle of the machine, and he realised that it was very inefficient because the cylinder was being repeatedly heated, cooled and then reheated as the steam was condensed.

He came up with a solution – the separate condenser. If the steam could be condensed outside the cylinder, the cylinder could be kept permanently hot. But that created a new problem. How do you keep a cylinder hot if the top of it is permanently open to the air? The obvious answer is to close it, but if you close the top off then you can no longer use air pressure to force the piston down.

Watt then made the great breakthrough – forget about air pressure, use steam under pressure to do the work. The atmospheric engine had become a genuine steam engine. This led directly to the next vital realisation. If you are going to use steam to push the cylinder, then you no longer have to rely on pump rods and gravity to move the beam in one direction. You could use steam on top of the piston to force it down and steam from below to push it up again. The engine could now become something far more important than a mere pumping machine. If you attached a device such as a crank to the end of the beam you could use it to turn a wheel. Now, there was a new and valuable power source that could be used to work the machines of factories and mills.

There were technical problems to overcome, some of which could be solved by Watt's own ingenuity, but others required outside help. The old atmospheric engines could be comparatively crude in construction, but for Watt's machine to work efficiently the parts had to be manufactured to far more exacting specifications. In particular, the piston had to fit tightly in the cylinder. The inventor had to turn to the ironmasters to supply the technology, and he found just the man he needed in John Wilkinson.

He was known as 'Iron-mad' Wilkinson with good reason, because of his obsession. It was an obsession that he carried with him throughout his life to the end, and beyond. He worked at an iron desk, slept in an iron bed and even had an iron coffin built for himself that, it is said, he kept propped up in a corner of his office; whether as an intimation of mortality or in the hope of attracting orders was never clear. At his death he was buried in the grounds of his own home, the spot marked by a 40ft-high iron obelisk. Unsurprisingly, a later purchaser of the house wanted neither Wilkinson nor his obelisk in his garden, so he and his monument were moved to a new resting place in what was then Cumberland, the county of his birth.

Wilkinson's father, Isaac, was the first of the family to enter the industry and soon established a reputation for innovative work, including the invention of a new type of water-powered blowing machine to provide the blast for the furnace. He formed a partnership to establish ironworks at Bersham near Wrexham in Wales, where he worked with both coke and charcoal furnaces. Isaac had raised the capital with the help of a Shrewsbury financier, Edward Blakeway. John Wilkinson was his eldest son and when his father moved to Bersham he stayed behind for a while in Cumbria. When he married an heiress, Ann Maudesley, in 1755, he and his wife moved to Bersham but she died the following year, after the birth of their daughter, Mary.

Although a partner at Bersham, he also became a partner in the New Willey Company, which had coke-fired blast furnaces near Broseley, just a few miles from Ironbridge. Blakeway was a partner in this enterprise as well as at Bersham, but he ran into financial difficulties and was declared bankrupt. His shares passed to his sister-in-law, Mary Lee, who married John Wilkinson in 1763. He was now in effective control of Bersham, but relations with his father were hardly cordial. Isaac sued John, and others, on the grounds that he was not receiving his fair share of the profits from his earlier work. Eventually, the affair was settled and Isaac left for new business interests in Bristol, leaving John in total control of the iron-making concerns.

John Wilkinson soon showed that he had inherited his father's inventive genius – and his litigious habits. On the inventive side, his first really important invention was for a way of boring iron guns from the solid. As explained earlier, iron guns were mainly made from wrought iron strips in an earlier period, and were not always reliable. James II of Scotland, for example, decided to offer a salute to his queen, lit the fuse on a cannon and was blown to pieces.

Wilkinson used a new technique. In previous bored gun barrels, the boring-bar was rotated as it advanced into the iron, but was liable to judder, creating an imperfect interior. He reversed the process, holding the boring-bar still and rigid, and rotating the barrel round it. The result was a much improved, accurately bored barrel. He patented the process, and although the navy tried to have the patent quashed in the national interest – claiming that it was quite wrong for one manufacturer to, in effect, have a monopoly on supplying ordnance – the court sided with Wilkinson.

The new bore may have been devised for artillery, but it was exactly what was needed for providing well-engineered cylinders for the new steam engines. Wilkinson was quick to see that an important market was opening up, and went on to develop a new type of cylinder lathe, which brought even greater improvements in accuracy. James Watt had an all-embracing patent for his new engine, and having formed a partnership with the Birmingham manufacturer Matthew Boulton, had a monopoly on steam engines. Some of the parts were manufactured at the Boulton & Watt Works, and these were supplied to the purchaser, who was expected to provide the rest, including the cylinder, and to erect the machine on site themselves.

Boulton visited Bersham in May 1775, and Wilkinson boasted that he could bore a cylinder from a solid block of iron in half the time it would take to cast it hollow. In fact it turned out to take twice as long but the result was a far better finish and Boulton was suitably impressed. He was so convinced of the superiority of Wilkinson's cylinders that Boulton & Watt insisted that his should be the only ones installed in their engines. It was an arrangement that was mutually beneficial – and Wilkinson, in turn, helped to provide a new use for the steam engine by using it to power his blowing cylinders for furnaces.

It was not, however, an entirely smooth relationship. Wilkinson had a prickly character and was certainly not mealy-mouthed. When James Watt objected to some criticisms made of his work, he laid into the engineer in no uncertain terms in a letter of 1779:

> Pray by what authority are you to be exempt from unmerited censure more than any others? And why cannot you submit to take the world as it is as well as any other of your friends. I thought myself one of the most unfit persons

> in the world to bear the haranguings of inconsiderate men but I am a hero in such circumstances compared with you … I thought you had been a philosopher but it appears you have got the failings of the clergy that you cannot practise what I have heard you preach.

Boulton & Watt's monopoly under the patent was constantly being threatened by engineers, especially in Cornwall, where they attempted to build engines to improve on Watt's design but which inevitably used some of the patented technology.

They were dubbed pirates, but there was another pirate nearer to home: John Wilkinson. He built engines for his own use without payment of any premium to Boulton & Watt. While still receiving preferential treatment as the sole provider of cylinders, he built a total of twenty-one engines for concerns in which he had a direct interest. He might have thought that was fair enough, but it would have been hard to deny a charge of piracy when he also built seven for other British customers and five for export.

All this might have remained hidden had he not quarrelled with his brother William, who felt he was not getting his fair share of the profits. This resulted eventually in a court case, with John Wilkinson threatening to close Bersham down altogether, but which ended with William being bought out, with the court awarding him £8,000 for dividend owed from past profits. The division had lasting effects. William turned away from John, and threw in his lot with Boulton. The manufacture of engine parts was now begun at the new Boulton & Watt Soho Foundry in Birmingham, and William helped entice key workers from Bersham to join the new enterprise. The partnership that had been so profitable to all parties came to an end with the opening of the Soho Foundry in 1796, but in its time it had been instrumental in creating a new form of power that was to dominate manufacturing and transport for another century, and more.

Wilkinson made another major contribution to technology, though at the time its significance was not immediately obvious. He took an idea that many thought absurd and put it into practice. It had actually been foretold as one of a series of prophecies made by Mother Shipton, who lived in a cave in the cliffs at Knaresborough, Yorkshire – now a tourist attraction. She said that men would fly, travel in horseless carriages and, perhaps looking forward to the internet, 'Around the world men's thoughts will fly. Quick as the twinkling of an eye'. In the sixteenth century when she wrote them down, these all seemed wildly unlikely, as did another of her ideas.

'In water, iron then shall float. As easy as a wooden boat.' This was the one that would be brought to reality by Wilkinson. Water transport, by river and canal, was essential to the success of his various enterprises, and he saw no reason why iron should not play its part here as well. Observation showed that if you put a piece of wood in water it floated; if you did the same with iron, it sank. He appreciated that if, however, you had a hollow iron container it would float because it was full of air.

He built his first iron barge, named appropriately *The Trial*, and launched it into the Severn at Coalbrookdale in July 1787. 'It answers all my expectations', he wrote, 'and it has convinced the unbelievers, who were 999 in a thousand.' Although development was slow, it was to mark the start of a fundamental change in the world of ship building in the next century.

The developments in iron making caused a major geographical shift in the industry. It was no longer necessary to build furnaces in areas such as the Weald, surrounded by woodland for charcoal making. Now furnaces needed to be close to coalfields, limestone and iron ore. New developments moved into South Wales, where all the essential ingredients were available. It was the industrialists from England who created the new landscape of the valleys – men such as Thomas Hill of Stafford who, with partners Thomas Hopkins and Benjamin Pratt, acquired a lease of land from the Earl of Abergavenny at Blaenavon.

Once again, the indefatigable tourists of the age provide us with a picture of the works in their prime. William Coxe visited the site in 1798 and described the scene in his *Historical Tour Through Monmouthshire* (1801):

> At some distance the works have the appearance of a small town, surrounded with heaps of ore, coal and limestone, and enlivened with all the bustle and activity of an opulent and increasing establishment. The view of the buildings, which are constructed in the excavations of the rocks, is extremely picturesque, and heightened by the volumes of black smoke emitted by the furnaces ... Although these works were only finished in 1789, three hundred and fifty men are employed.

Something of that grandeur remains, including the ruined blast furnaces and a very imposing structure, the water balance tower. As Coxe explained, the site is built into a hillside, and the tower provided a means of getting material up and down between the two levels. Basically it consisted of two connected containers, each fitted with a water tank. The balance was achieved by adding water to the tank on one container and draining it from the other, the weight of the one with the full tank going down being used to pull the other up.

The enterprise was connected to limestone quarries by a tramway, while another led away to connect with the Brecon & Abergavenny Canal, which in turn joined the Monmouthshire Canal that connected to the rapidly growing port of Newport. There are the remains of five furnaces, together with the casting houses at their foot. One can see the advantages of the site. Being set into the hill, there is direct access at the top for loading fuel, ore and stone, which was originally stored in the charge house, while the molten metal can be tapped at the lower level. A stack indicates the site of the original blowing engine. The location may have been remote, set in wild countryside, but it was not isolated. From Newport the iron could be shipped anywhere.

The English ironmasters who came here faced an immediate problem. Near the works was a tiny hamlet, and surrounding that nothing but bleak moorland. Yet when Coxe came here just ten years after the works had opened for business he found 350 men at work. One of the first and most pressing needs the company faced was having to provide somewhere in which to house them. They came up with some ingenious ideas.

Part of the tramway complex serving the works passed over a viaduct carried on ten arches. These were bricked in and given doors and windows to supply basic accommodation. Coxe described the scene at the viaduct: 'Numerous workmen continually pass and repass, and low cars, laden with coal or iron ore, roll along with their broad and grooved wheels.' He described it as being 'a singular and animated picture'. It must also have been an extremely noisy place, and could hardly have been very peaceful for anyone living underneath this simple railway.

At the nearby Dowlais Works at Merthyr Tydfil an even more radical piece of infilling was tried. This time, the arch bricked in to provide a primitive house was that of the bridge joining the bank to the top of the furnace. This turned out not just to be uncomfortable, but fatal, as described in a newspaper report of June 1793:

> Old Edward Maddy … his wife, and another old Man, found dead in their House under the Bridge House in the Old Furnace, Suffocated as it is supposed (and without doubt it is so) by the Damp coming thro' the Air Holes of the Furnace into their House.

The damp referred to was not moisture, but fire damp, the name then used for carbon monoxide. The unfortunates were poisoned by the fumes.

Living quarters such as these were hardly likely to attract skilled men, many of whom were recruited from England, to come and work in South Wales. We can still see the solution at Blaenavon, where an open square of houses has survived. They were not grand – the average floor area is a mere 19ft by 14ft – but they were solidly built of rough sandstone, with stone flagstones on the ground floor. A map of the site, dated 1819, shows the square much as it is today, with one of the houses being used as a shop. This would have been run by the company, a system that could easily be abused. The workers and their families would have had no choice and would have had to pay whatever the company decided for their most basic provisions, but on the other hand there was little other option in what was still a totally undeveloped part of the country.

Run fairly, the system worked well for both parties. In the early part of the nineteenth century, the company built more, mainly three-bedroomed, houses in Blaenavon. Other ironworks had similar problems and built houses for the new workforce. At the Ynysfach Works built by the Crawshay family, for example, they provided terraces of cottages, all quite small and even smaller than those at Blaenavon, but equally solidly constructed.

The Blaenavon Ironworks in Wales in 1800; casting houses stand in front of the tall blast furnaces.

The living space was cramped and dark – the first floor windows are tiny and right under the eaves, so they can never have let in very much light. They were built close to the furnaces, which might have been convenient for getting to work but at the same time exposed the houses to all the fumes and dust that are an inevitable part of the iron-making process.

That so many early houses have survived in the area is a tribute to the fact that they were sturdily built, unlike some of the slum properties that were thrown up in other burgeoning industrial areas. It is interesting, however, to compare these houses with the Crawshay family home, Cyfarthfa Castle, a grandiloquent Gothic mansion set in magnificent parkland. It is a sign of the great wealth that the ironmasters of South Wales enjoyed.

There is more evidence of this in the Trevithick story. It was Samuel Homfray, of the Penydarren Works at Merthyr, who had invited the engineer to build a machine that would be capable of acting both as a locomotive to haul trucks on his tramway and as a stationary engine to provide power at the works. Richard Crawshay thought the whole notion absurd. He bet Homfray 500 guineas that the engine would not be able to pull 10 tons of iron down the tramway and return with the empty trucks. In February 1804 Homfray won the wager, but what is remarkable is the sum of money involved – that would be an £18,000 bet today. These were seriously rich men.

Decent housing, at least by the standards of the time, was just one incentive offered to lure good workmen to Wales, but sometimes that alone was not enough and there were times of serious labour shortages. When that happened, the ironmasters were none too fussy about the methods they used to fill their vacancies. One of the grandest of all the companies was Dowlais, but even they were not above a little attempted poaching, as this letter to Joseph Guest of Dowlais from a rival, Joseph Priest, indicates: 'I conceive thou canst not be unaware that someone from your works has been here twice enticing our men away by offers of high wages.'

It was, however, one thing to offer higher wages as a lure when the employers were desperate to secure extra labour, quite another when the men took it upon themselves to try and move to somewhere with better pay. The employers closed ranks. In 1803, William Wood of Penydarren wrote to his fellow employers:

> In consequence of a disagreement with some of our Pudlers [*sic*] we are apprehensive that the following Men or some of them may leave the place

> and in that case apply to you for work. Should they make this application we hope and request that you will not employ them, the dispute being such as materially concerns every iron-master in the Country.

The arguments over pay went backwards and forwards, with ironmasters constantly complaining about outrageous wages paid by competitors, while everyone tried to keep the bills down at the same time by cutting pay. It was a circle they could never quite square.

Richard Crawshay wrote indignantly in May 1797 about his attempts to change the piece-rate system, under which he declared men were getting 'such excessive wages as are Scandalous to pay', and replace it with a weekly wage. The men turned him down flat, telling him that the unskilled cinder-wheelers, basically the men who removed the slag, were able to earn 15*s* (around 75p) a week, with house and fuel provided, at Dowlais. Such largesse horrified Crawshay, who continued his letter, 'if when any of our Men or yours insist upon Wages incompatible the other will Countenance them by immediate employ, we shall injure all our Works and make resistance to all reasonable remonstrances with the Workmen in vain'. It is worth considering for a moment that this is the man prepared to lay out 500 guineas in a single bet, enough money to pay over a dozen men's wages for a year at that 'outrageous' rate of 15*s* a week!

Workers were hampered in their attempts to get better pay by the various Combination Acts that appeared from 1799. The first Act made it illegal for workers to get together to push either for a rise in wages or a reduction in hours, an offence that was punishable by three months in jail or two months' hard labour. Similar punishments were imposed on those who 'by any Means whatsoever, directly or indirectly, decoy, persuade, solicit, intimidate, influence or prevail, or attempt to prevail' on anyone to leave their work as part of an industrial dispute. Workers were also forbidden to put any money aside to help pay for attempts to improve their conditions. They were, in effect, being told that it was their place to accept the pay that employers chose to offer them.

It was true, an individual could ask for a rise on his own behalf, but the chances of a single individual being able to negotiate a payment purely on his own were on a scale from negligible to non-existent. Ironically, it was the man famous for fighting to abolish the slave trade abroad, William Wilberforce, who was one of the most passionate advocates of outlawing what he called the 'general disease' of combinations of workers. The masters of South Wales were equally vocal in their opposition to all forms of combinations as being entirely opposed to their fundamental principles of freedom, competition and enterprise – except when it came to their own interests. In January 1811, the ironmasters of the region all met to fix the prices for all kinds of tramway rails.

Often the men who had come to the region were under contracts that held them to strict working conditions, but there was nothing in the agreements that said the masters could not change those conditions themselves when it suited their interests. In 1799 the workers at Dowlais complained that the price of the coal they bought from the company for their own use had been increased without any consultation. One unfortunate worker was given the task of asking the company to change their mind. But he was acting for others, against the terms of the new laws, and was promptly jailed. He wrote a sad, misspelled note to the master, 'I ham sorry that I abuse your Honor in taking so much Upon me to Speek for Others' and asked to be set free. There is no record of whether his plea succeeded.

Other workers who tried to complain about a decision could fare even worse. The owners had real sanctions they could apply – a man who was sacked for complaining would also be evicted from the company-owned home occupied by his family. This is what happened to others in 1799.

The Dowlais Company had been paying a bonus of a guinea when production rose above an agreed level. That year, they discontinued the payment on the grounds that the improvement was nothing to do with a greater effort being made by the workers, but was entirely down to investment in improving the blast. The workers objected and the spokespersons were dismissed. One of them wrote at length, explaining his case and how he came to be one of the two men who spoke for the rest:

> About 18 months back I agreed to Come to Dowlais to be keeper at No.3. So when I went to set to work I found the furnace in very bad Condition for the hearth would not hold the iron of 3 hours blowing and Mr Onnions tould Me Several times that they Could do no good with her from the first blowing in to that time I came to her, but however I got the furnace in good Condition Enough in about the Space of 9 days or a fortnight and Sir I refer you to the truth of what I write to Mr Onnions who I make no doubt will Sertify the same: I did not Expect no reward, but I must confess I thought I should have been placed a little more in the Confidence of my Employers than if the furnace had done well before, all this I submit to your

> Consideration: and Now Sir another thing I have to Lay before you is: we made at No.3 Something above 51 tons of iron about 3 weeks back and the other two furnaces had made something above 40 tons each. So Dick Davis happened to go the office first and the guenea was refused to him as was Costomary, so he Came and told the Rest of the keepers and me how it was. Then they all declared that they would not work Except they should have it. So we all went together to the office, and because Dick Davis and me Could Speak English they deseired us to Taugh for them as well as our selves. So Consiquently there was Some dispute but there Was nothing spoke that was vexatious. But however we insisted on having the guineas that was then due, and we did not Look at it to be just to stop this money without any previous Notice; and then we were Willing to work, for the same as they did in other iron-works, or we would Come to aney other agreement that was reasonable; and now Sir I am informed by Mr Onnions that I am to be discharged and Dick Davis likewise because we spoke and the rest did not, when at the same time they was all there and spoke the words in the Welsh tongue to us as we spoke to the masters …

Again, the answer to the plea has not been preserved. For many of the owners, the matter was very simple and Thomas Guest of Dowlais even claimed that his rights as an owner were divinely ordained. Asking for higher wages was not merely unacceptable to authority but a sin:

> In providing for your own house you are not to infringe on the providential order of God, by invading the rights of others, by attempting to force upon those whom God has set over you, the adoption of such regulations and the payment of such wages as would be beneficial to yourselves.

It is an idea repeated in the popular nineteenth-century hymn 'All Things Bright and Beautiful':

> The rich man in his castle,
> The poor man at his gate,
> God made them high and lowly
> And ordered their estate.

The hymn is still sung but, not surprisingly perhaps, that verse is now omitted from modern hymnals.

There was another cause of complaint among the men: the prevalence of two closely allied systems, truck and the tommy shop. In the former, the men were paid partly in goods instead of cash, and in the latter they were obliged to buy their provisions in the company store, the tommy shop.

Various Truck Acts were passed by Parliament. The worst effects were felt in the mining districts, some of which coincided with the iron-making areas – and often the same employers ran both mines and foundries. At Tipton in the Black Country in 1822, the masters decided to counteract a fall in the price of iron by dropping wages and at the same time raising prices in the tommy shop, even though by this date the practice was illegal. A correspondent to the Home Office gave details of what was happening:

> The Coal and Iron masters compel their workmen to accept of two-thirds of their wages in goods such as Sugar, Soap, Candles, Meat, Bacon, Flour, etc., instead of money, at an unreasonable large profit. This appears to be the real cause of complaint rather than the reduction of wages, and is really very hard upon them, and as the masters contrive to evade the Act of Parliament the men seem to have no relief but ceasing to work.

Faced with a threat of strike action or worse, the magistrates stepped in to enforce the law – not, however, the law specifically banning the truck and tommy systems. They decided that the men were actually being treated with great fairness, but they announced they intended to rigorously enforce the vagrancy laws against any men who tried to beg for money while they were out on strike.

A similar battle was fought across the border in Wales in 1832 at Dowlais where, it was reported, the men were paying an average of 15–20 per cent above the normal rate for goods they were forced to buy from the company. That same year, however, at a meeting of masters at Abergavenny, agreement was reached to abolish the system altogether. It seems extraordinary that a special meeting had to be called to get the masters to agree not to break the law of the land.

The troubles in the Staffordshire area were due, in part, to the way in which the industry was organised. A letter of 1819, from an anonymous writer from Shropshire, cast a jaundiced eye over the works of the neighbouring county:

> In Staffordshire landed property is very much divided; and, naturally, all the proprietors desirous of turning their coal and iron mines to *immediate* account. Hence there is a colliery in almost every field. As there is not sale for such an immense quantity of coal and ironstone, several of those little proprietors unite together and build furnaces; clerks from the neighbouring manufactories are taken in as partners to direct the concerns; the tradesmen of the towns in the vicinity who can raise a hundred or two hundred pounds, form part of the firm; and it is in this way that the iron-works have been multiplied in that county. The proprietors embark all their property, and *all they can borrow* in these establishments. This slippery foundation is rendered still more so by an inferiority in the quality of their iron when compared with that of South Wales and Shropshire. From this latter circumstance, whenever the make of iron exceeds the demand Staffordshire is the first in feeling the deficiency of the orders, and as the trade of that county cannot (for want of capital) bear stock, they immediately reduce the price to obtain new customers.

So, in Staffordshire, the only way to compete was by undercutting other areas such as South Wales, and the only way to do that was to cut the wage bill. It is no wonder that the writer refers to Staffordshire as 'a millstone round the neck of the iron trade'.

Coke smelting in Scotland began with the foundation of the Carron Company near Falkirk in 1759. It was here that James Watt worked with one of the company's founders, Dr John Roebuck, to develop his steam engine, before moving down to Birmingham to join Matthew Boulton. One of its most famous products was a short cast-iron gun, the carronade. It was greatly admired by military specialists and was installed on Nelson's flagship the *Victory* and used by Wellington at Waterloo. This was only one of the products of this pioneering Scottish ironworks.

From South Wales to Scotland, there was a great proliferation of furnaces at the end of the eighteenth and beginning of the nineteenth centuries. Both cast and wrought iron was being produced in previously undreamed of quantities. As more iron was produced, so more and more new uses were being found, old industries were transformed and demand continued to grow.

4

NAILS AND CHAINS

Nails and chains had been made since antiquity, but with the arrival of new methods of making iron the manufacturing processes were changed.

Originally nails were made by the local blacksmith, but by the seventeenth century it had become a specialised trade, centred mainly in the Black Country, in Dudley and Wolverhampton. It developed as a cottage industry, carried out in small workshops, often little better than crude lean-tos tacked on to the side of a house or sheds in a backyard.

The equipment needed was quite simple, based on the 'oliver'. The hammer head used for forging was attached to a flexible wooden pole, but by pressing on a foot pedal the hammer head could be brought down with considerable force on to the metal on the anvil. As soon as the foot was lifted from the pedal, the pole sprang back into position, raising the hammer. The nailer would generally operate two hammers, set close together. The wrought iron was heated in a hearth next to the nailer, who would pick up the length of iron and place it on the anvil, which had been specially prepared. This was generally made of stone, but covered with a metal plate, with shaping devices set into it to create the nails. The task was simple, repetitive and poorly paid.

The nailmaster provided the iron rods and also bought the finished nails. They controlled the whole system and were notoriously unscrupulous. The worst of them were known as 'foggers'. The nailers bought iron rods from the foggers, paying by weight, and then sold the finished nails back to them, again being paid by weight. It was widely believed that the foggers kept two sets of scales: one for the iron they sold and the other for the nails they bought. The two, it was said, never agreed and no one was in any doubt about who the differences favoured.

Forging chains by hand in the Black Country.

There were further abuses in the system. Many of the foggers paid in tokens instead of coins of the realm, tokens that could only be exchanged in their own provision shops. The nailers complained bitterly about their treatment and the many devices used to cheat them, which included selling them inferior iron that could not be made into nails, but they had no redress. No one could afford to tell one of the nail merchants that they would no longer work for them again, as there were too many poor families willing to take their place. And this was a family concern, a great deal of the work being done by women and even children. It has been estimated that by 1830 there were 50,000 working in the trade, with, it seemed, every house in areas such as Dudley having a workshop in the backyard. As all of them were belching out smoke from the furnaces it is easy to see how the area became known as the Black Country.

Living conditions for nailers deteriorated through much of the nineteenth century, as the competition from machine-made nails increased and the price of iron slumped. A bundle of iron rods, or spikes, could be bought by the nailer and if the iron was good he could make a maximum of 10*s* a bundle. A family could reasonably expect to make up ten bundles a week into nails and, although they were highly unlikely to get the highest rate for everything they made, they could still expect to clear £3–4 a week in the 1830s.

Fifty years later, a family would find the price had fallen to just 8*d* a bundle, and they could only survive by parish relief. They were handed meal tickets, which entitled them to a loaf of bread and a quart of soup. The only part of the trade that continued to pay anything even approaching a living wage was the making of nails for horseshoes, the one part of the trade that had not been mechanised.

The law did little, if anything, to help the plight of the nailers. In spite of legislation, the tommy shop system continued until nearly the end of the nineteenth century. A commission of 1882 found it was still prevalent. Even after the repeal of the various combination laws, the nailers were no better off. There was no organisation, simply because the individual families largely worked on their own and made their own deals with the merchants. There were attempts to organise strikes, that had little effect, and some led inevitably to riots that were suppressed with the help of local militia. Some nailers turned to the other great trade of the area, which was believed to pay rather better. They became chain makers. It did not necessarily make for any great improvement in their lives.

At the beginning of the nineteenth century there were two important chain-making centres: Pontypridd, where the Brown Lennox Works supplied chains to the navy and the merchant marine (a company we shall hear more of in the next chapter), and the Black Country, especially Cradley Heath, where a whole number of small works sprang up.

Throughout the eighteenth century, ships' anchors had been suspended from heavy manila rope cables. The problem with replacing these with iron chains was that wrought iron tended to stretch when put under strain. The difficulty was solved by Thomas Bunton, who set a stud in the centre of each link. It was a simple technique. A finished link was heated to a white heat, and the cold stud placed in the centre. As the link cooled, it contracted, tightly gripping the stud.

The first ship's cable was said to have been forged by Noah Hingley. He had begun his working life running a small chain-making business and forge. He expanded his interests when he leased land from the Earl of Dudley, which gave him access to coal, iron ore and limestone, all he needed to start up in business as an ironmaster. His first ship's cable, forged by hand, was made in 1820 and, once success was ensured, he established a much bigger chain-making factory at Netherton in 1837. It was to become the biggest chain-making company in the world and it added another essential component to its business – it also forged anchors. Its biggest ever task was to make the anchors for the ill-fated *Titanic.* These massive anchors were each 18ft 6in long and weighed 15 tons 6cwt. Amazingly, they were forged by hand and at the time they were the biggest ever made.

Unlike nail making, heavy chain manufacturing took place in factories, large or small. A typical chain shop consisted of a series of hearths set around the edge of a central area where the links were forged. The hearths were made of brick and held a coke fire, which was brought to high heat by the use of bellows, usually operated either by a foot treadle or an overhead beam. The iron pipe that blew the air into the fire had to be water-cooled to prevent it melting in the intense heat. It was this heat that made working in a chain shop so arduous. Men wore aprons made of coarse tarpaulin or some other dense, heavy material for protection from the fire, and had to drink water throughout the day to prevent dehydration: these were, quite literally, sweatshops.

The process began with cutting the hot metal into appropriate lengths using a guillotine. The heated rod had first to be bent into a U-shape over a former designed to give it the correct shape, and once that had been achieved, the bent rod could then be slipped through the last completed link of the chain, then bent again and the ends welded together by hammering on the anvil. With heavy anchor chains, which could be 1½in, or even more, in diameter, this process required a whole team of men working together. The smith would control the process, holding the piece on the anvil, while two or more strikers hammered it into shape and welded the ends together. The weld was completed on a tommy, a device very like a bigger version of the nail maker's 'oliver'. This also required a springy beam, worked by foot treadle, to power the hammer.

Smaller chains were forged by hand, often in workshops at the backs of houses, and much of this work was undertaken by women. They were notoriously badly paid and conditions were harsh. They suffered the same abuses through tommy shops and wretched conditions as the nailers. This system continued right up to the early years of the twentieth century, when this and other forms of sweated labour finally attracted public attention.

In 1909 the Liberal Government passed the Trade Board Act to set up official bodies to regulate minimum pay for four categories of workers, chain makers among them. They decided that the minimum rate for the chain makers working by hand should be set at 2½*d* an hour (roughly

60p an hour at today's prices). In other words they would have to work for ninety-six hours to earn £1 – a huge improvement on the rates most of them had been getting up to then. In August 1910, the majority of the employers simply refused to pay the increase, and there was nothing in the legislation to enforce the recommendation of the Trade Board.

At this point, the women of Cradley Heath took matters into their own hands. In the previous century, the women would have had little chance to fight for the increase, but the Combination Acts had been repealed, a new Trade Union movement was growing and women had developed their own organisations.

One of the pioneers was Mary Macarthur. She was an extraordinary woman who was born in Glasgow in 1880. Her father was a draper and she was brought up in a typically conservative household, even joining the local Primrose League, named after the favourite flower of the former Prime Minister Benjamin Disraeli. She might have followed the conventional path of respectable middle-class life if her father had not heard about a meeting that had been called to form an Ayrshire branch of the Shop Assistants' Union. He sent his daughter along to infiltrate the meeting to find out what was going on. The result was not exactly what he had anticipated.

Mary was totally won over by the arguments and, instead of helping her father to oppose the union, she joined it. She was soon chairing meetings, and became the first woman to be given a seat on the National Executive. Shortly after that, she moved to London to take on the job of secretary to the Women's Trade Union League. In 1906 she founded the National Federation of Women Workers. She now decided to give full backing to the women of Cradley Heath, offering strike pay to her union members and even providing funds, though at a reduced rate, to non-members.

The strike began in August 1910, and the union was able to make a powerful case for action simply by using all the official figures and facts that had appeared in the Board of Trade report on conditions in the industry. A typical case was that of Miss A., who was signed in as an apprentice at the age of 13 and bound for two years. After just three weeks, she was set to chain making at a wage of 2*s* (10p) a week that rose to 3*s* 6*d* a week (around 17.5p) after a year. She was required to make 600 links a day, working from 5 a.m. to 5.30 p.m. Once her apprenticeship was ended, her pay might be expected to rise to 5*s* a week.

Another woman was reported as having made 2cwt (100kg) of chain a week. The report also gave information on living conditions. A typical chain maker's house had a living room and scullery on the first floor, with two bedrooms above and a washhouse in the yard, for which the usual rent was 1*s* 6*d* a week and an extra 3*d* a week was paid for renting the hearth.

The press took up the story of the hard-pressed women, including this account of an elderly worker that appeared in the *Daily Express* on 1 September:

> The most pathetic figure in the strike of women chain makers at Cradley Heath is Mrs Patience Round, an old woman of seventy-nine, who has raised her feeble voice with the other women who are demanding the right to live.
>
> It is now sixty-seven years since Mrs Round, as a young girl, started on her long years of chain making. Since that day the world has been the forge in her backyard. The great happenings in the world outside have never penetrated the smoke begrimed world of her home, where day after day and year after year she has ceaselessly beaten the glowing iron into shape and worked the bellows until her figure has become bent and her hands indented with the marks of the chains she has forged.
>
> After sixty-nine years the old woman has laid down her hammer and bravely defied her employers by joining others who demanded wages of 2½*d* an hour instead of 1½*d* an hour.
>
> She rose at 4 in the morning and worked till the light faded.

These and similar reports roused public opinion and brought in nearly £4,000 in donations to swell the fighting fund. The women even had a special song written for them, 'Rouse Ye Women', sung to the tune of 'Men of Harlech'. Here is the chorus and first verse:

> Rouse ye women, long enduring,
> Beat no iron, blow no bellows
> Till ye win the fight ensuring
> Pay that is your due.
>
> Through years uncomplaining
> Hope and strength were waning –
> Your industry
> A beggar's fee
> And meagre fare
> Were gaining.
> Now a Trade Board is created,

See your pain and dearth abated
And the sweater's wiles checkmated
By Parliament's decree.

In the first week of the strike, 414 women laid down their tools, but by the second week the number of strikers had grown to 656, at least two-thirds of the total workforce. Within a month two-thirds of the employers had capitulated and agreed to the new minimum wage, and by the end of October all had signed up. The strike was over and the women had won.

Perhaps the most surprising thing about these two industries is that they retained their old craft systems for so long, with only the most basic form of mechanical assistance to help with the brute force of muscle power. One reason must be that it suited the employers, who had no incentive to invest in expensive machinery when they could rely on keeping wages low to ensure their own profits. In other parts of Britain, new uses of iron were being found and developed that would produce profound changes and would have widespread effects.

5

THE 'COLOSSUS OF ROADS'

This punning name was given to the great civil engineer Thomas Telford by his friend, the poet laureate Robert Southey. He was to have an immense impact on the iron industry when he used the material in a whole range of different, pioneering structures for the rapidly developing transport system of the late eighteenth and early nineteenth centuries – roads and canals.

Perhaps the most surprising thing about Telford is that he ever became an engineer at all, let alone one of, if not the, greatest of the age. He was born in the Scottish Lowlands in Eskdale in 1757, where his father was a shepherd. It was a happy occasion for the family, for their first son, also called Thomas, died in infancy. Their delight was cruelly short-lived. Just two months after his birth, Thomas' father died, and he and his mother were forced to move to an even smaller cottage in the same valley, where they rented a single room.

His mother, Janet, was to live there all her life, a proud woman who refused to accept a better home, even when her son was becoming famous and comparatively wealthy. But with the loss of her husband, she was forced to accept help from her brother, and young Thomas eventually went to the local village school to receive a very elementary education. He went on to learn a trade, being helped by the family to take an apprenticeship to a local stonemason.

There must have been something special about the young man, for he attracted the attention of one of the wealthier families in the area, the Pasleys, especially Elizabeth Pasley, who was always keen to share her passion for literature with anyone who might benefit from her enthusiasm. Young

Telford was to be her star pupil, and in later life he recalled the first book he read that grabbed his attention. Not, perhaps, the likeliest reading matter for a young man, it was Milton's epic poem *Paradise Lost*. Telford was never to lose his passion for poetry, and he was to write verse himself throughout his life. In one verse, published in *Ruddiman's Edinburgh Magazine* in 1779, he wrote words that might well have been a summary of his own life:

> Nor pass the tentie [attentive] curious lad,
> Who o'er the ingle hangs his head,
> And begs of neighbours books to read;
> For hence arise
> Thy country's sons, who far are spread
> Baith bold and wise.

Telford proved himself to be more than a merely competent mason, and throughout his life he prided himself on his craftsmanship. Even when he had become established as an engineer, he never forgot where he had started. According to the biography written by Samuel Smiles in 1862, he met an old friend many years later, who had given up hacking stone to take the less arduous job of innkeeper. Telford asked him if he still had his old tools and was told he had lost them long ago. 'I have taken better care of mine,' said Telford. 'I have them all locked up in a room at Shrewsbury, as well as my old working clothes and leather apron: you know one can never tell what may happen.'

There is still evidence that Telford retained his old skills in the churchyard at Bentpath, near his birthplace, where he returned in later life to carve the memorial stone to his father that they could not afford at the time he died. He also carved a monument to the Pasley family, not just to Elizabeth, but also to John Pasley, to whom he was also greatly indebted for helping him advance in the world.

Competence alone is not always enough to give a young man a chance to prove himself. Sometimes a helping hand from powerful friends can make all the difference. In Telford's case, there is no doubt that his natural intelligence and love of literature helped to make him seem worthy of the attention of families such as the Pasleys. With their help, he was able to move out of the narrow restrictions of life in Eskdale to see the new world that was emerging elsewhere. He had gone on his own to Edinburgh, where he had been able to find work on the new town that was being built in the shadow of the castle. There he discovered a very different environment, where a new style of neo-classical architecture was being developed that, as much as anything, has come to define the Georgian age.

In 1781 he was back in Eskdale – but not for long. Soon he was on his way to London where, thanks to John Pasley, he got introductions to two of the leading architects of the day, Robert Adam and Sir William Chambers. Soon he found himself at work on one of the most prestigious new buildings in the capital, Somerset House.

Telford soon realised that he was more skilled than the average London mason – and a great deal more ambitious. While working on Somerset House he met a workman called Hatton, who was remarkably skilful at intricate carvings in stone and marble, but was working for a master who treated him like an ordinary workman. Telford planned to persuade him to go into partnership together, offering specialist skills to the leading architects of the day. In a letter home to his old friend Andrew Little, he described his plans with all the cocksure enthusiasm of the young and confident: 'The Master he works under looks on him as the principal support of his business but I'll tear away that Pillar if my scheme succeeds, and let the Old beef head and his puppy of an ignorant Clerk try their dexterity at their leisure.'

The partnership never happened, but Telford was soon making new connections with the powerful and influential. It is not necessary to list all the different schemes he worked on as he steadily ascended the ladder of success, which included an important job working on large-scale improvements at the Royal Navy dockyard at Portsmouth. All the time, he was embarking on an ambitious programme of self-education in what little spare time he had:

> Knowledge is my most ardent pursuit, a thousand things occur that would pass unnoticed by good easy people who are contented with trudging on in the beaten Path but I am not contented unless I can reason on every particular. I am now very deep in Chemistry – the manner of making Mortar led me to enquire into the nature of Lime &c. in pursuit of this, having look'd in some books on Chemistry I perceived that the field was boundless – and that to assign reasons for many Mechanical processes it required general knowledge of the Science. I have therefore had the loan of a M.S.S. Copy of Dr Blacks Lectures. I have bought his experiments On Magnesia And Quick Lime and likewise Fourcroy's Lectures translated from the French by a Mr Elliott, Edinburgh. And I am determined to study with unwearied attention until I attain some general knowledge of Chemistry as it is of Universal use in

> the Arts as well as in Medicine. I wish Andw that you saw me at the present instance surrounded by Books, Drawings – Compasses, Pencils and Pens etc. etc. great is the confusion but it pleases my taste and *that's enough.*

Among the influential men he met was Sir William Pulteney, who was the MP for Shrewsbury and whose estates included the very dilapidated Shrewsbury Castle. Sir William decided that it needed to be transformed into a suitable home for a gentleman and offered the job of architect and overseer of works to Telford. In 1786 he set off for Shropshire, in what was to prove the decisive turning point in his life.

He had a very successful time in Shrewsbury, mixing in society, interested in the world about him, but steering clear of local politics. He even went to work, using local convicts as a labour force, to excavate the Roman remains at nearby Wroxeter. He obviously made a good impression on the people who mattered, for he was invited to take up the post of county surveyor for Shropshire. There were no special requirements for the job, and many simply regarded it as a sinecure, but Telford took it seriously.

He now had the opportunity to act as an architect in his own right, instead of a builder following the plans prepared by others. He built two churches in the county, at Madeley and Bridgnorth, both in a rather severe classical style. They are competent pieces of work, but hardly more than that. He could probably have managed to make a living as a local architect with a restricted practice. He also had an opportunity to practise his old craft, when he built a bridge across the Severn at Montford, which was entirely his own design. Again it was a perfectly respectable work, built on three sandstone arches. But another opportunity now came along that was to set him in a very different and, as it turned out, far more promising direction.

The 1790s were the years of 'canal mania'. Following the success of the pioneering canal built for the Duke of Bridgewater, opened in 1761, there had been a flurry of canal construction. The chief engineer for most of the new projects was James Brindley, but when he died in 1772, a new generation of engineers took over, led by William Jessop.

After a lull in the 1780s, construction began in earnest and 1793 was a particularly hectic year, with Parliament giving their approval to eighteen new Canal Acts. Jessop was in huge demand and was appointed chief engineer to several projects, including the hugely important Grand Junction that was to link London and Birmingham. He was also in charge of the rather less significant Ellesmere Canal, begun that same year. This was to link the Chester Canal to ironworks in North Wales, on the far side of the Dee Valley.

Jessop laid out the basic plans, but he could not be everywhere so he needed an experienced, trusted man on the spot. He had hoped to have one of his own assistants take on the job, but the committee did not even consult him. They offered it instead to their new local hero, Thomas Telford. He was cock-a-hoop and wrote to his friend, Little, on 29 September 1793, 'last Monday I was appointed Sole Agent, Architect and Engineer to the Canal which is to join the Mersey, the Dee and the Severn. It is the greatest work, I believe, that is now in hand, in this kingdom, and will not be completed for many years to come.'

It was not the greatest work in the kingdom, nor was he the sole agent – Jessop was still the man who ultimately took all the decisions – but it was an opportunity to move into a new and vibrant world. As an architect, he had simply followed trends; now as an engineer he found he had the opportunity for innovation. It came about, literally, by accident. He had scarcely been in the job for a year when he found himself elevated to the rank of chief engineer for the modest Shrewsbury Canal, an 18-mile-long waterway linking Shrewsbury to the Ketley Canal and the old Shropshire Canal.

The Ketley Canal was the more interesting, for instead of the usual flight of locks, joining one level of the canal to the other, it had an inclined plane. In this system, a simple tub boat entered a lock at the top of the plane where, as the water was let out, it settled onto a wheeled carriage running on rails. As the main traffic on the canal was from the ironworks at the upper level, most of the cargo was downhill and the weight of the boat going down could be used to haul up an empty tub – a system similar to that of the water balance tower at Blaenavon. The man behind this ingenious device was William Reynolds, the ironmaster whose work in Coalbrookdale we looked at earlier. There was only one outstanding engineering problem on the new Shrewsbury Canal, one that was easily solved – but Reynolds found in Telford a man who shared his enthusiasm for trying out new ideas.

Telford had the job because the original chief engineer, Josiah Clowes, had died suddenly. He had already begun work on a conventional aqueduct to carry the canal over the River Tern at Longdon. There was no need to provide anything very special as the river was not navigable and the job was straightforward, but when Telford arrived on the scene he found that Clowes' work had all been swept away in a flood, leaving only the stone abutments on either bank.

The two men decided not to rebuild the proposed masonry aqueduct but to try something entirely novel. The canal would be carried in a wrought-iron trough, supported by iron struts. The canal has now gone, but the aqueduct itself has survived. The contrast between the lightweight trough and the massive stone abutments is striking. The trough consists of twenty-six sections bolted together, with a towpath slung on the outside of the trough. These are supported on three inverted iron triangles set into masonry bases. Not only is the iron trough lighter than a stone one would have been, but the weight is further reduced because the iron is watertight. The porous stone would have to have been lined by heavy 'puddle', a gooey mixture of clay, sand and water.

If that was all there is to the story, it might not be significant – it is not even certain that this was the first iron trough aqueduct. Another engineer, Benjamin Outram, was constructing a similar aqueduct on the Derby Canal, but that has long since been dismantled. Outram was partner in the Butterley Ironworks, and he and Reynolds were probably moved to experiment by the same motive. It was not that they necessarily thought iron aqueducts superior, but they provided more work for their foundries. The Longdon aqueduct is important because of what happened next.

Back at the Ellesmere Canal, Jessop was wrestling with a very difficult problem: how to take his canal across the deep valley of the Dee near Llangollen. The obvious solution was simply to build a multi-arched aqueduct across the valley. The trouble was that from the point where the canal arrived at the rim of the valley, the aqueduct would need to have been some 2,500ft long to reach the other side and would stand over 100ft above the waters of the Dee.

One solution was to follow a design first used by the Romans for their aqueducts – carry the trough on a series of double arches – but this would have been exorbitantly expensive. Drawings for this type of construction were made, but were never executed. The next alternative was to build an aqueduct at a lower level, but this would have involved flights of locks at either end to bring the canal down one side of the valley and lift it up again at the other. The problem was that the piers would have had to be massive to support the weight of the long masonry trough and its lining of puddle. The arrangement also created another problem. The trough would have acted as a sump, with water draining down into it, from both ends, which would then have to be pumped back up to the canal: yet more expense. But now, Telford was back from Shrewsbury with the answer: a light iron trough to cross the entire span.

We do not know how much thought went into coming up with the eventual solution to the Dee crossing. There is one unlikely sketch, signed by Telford, which shows the trough supported on spindly, lattice columns of iron.

The final design was rather more conventional, but was still remarkably daring. It was decided to avoid any locks at the ends, which meant raising the level of the canal to the south of the aqueduct by means of an embankment, in order to be at the same level as the opposite bank. When it was completed, the trough was 1,007ft long and rose 121ft above the Dee. It was carried on nineteen arches, each built of solid masonry to a height of 70ft, above which, in order to decrease the load, they were hollow and braced with cross walls. The trough itself consisted of an elaborate system of iron plates bolted together, supported by iron ribs, with the towpath now inside the trough and cantilevered out over the water. The ironwork was supplied from the nearby Plas Kynaston Ironworks, run by William Hazledine, and Telford was so impressed by the workmanship that he christened him 'Merlin' Hazledine.

During construction, Jessop was concerned that the masons working on the piers would be 'giddy and terrifyed' working at such heights, but Telford worked out a construction plan that solved the problem, as he described in his autobiography, published in 1838:

> The stone piers ... were all built to the level of 20 feet, and then the scaffolding and gangways were all raised to that level, and the materials being brought from the north bank; the workmen always commenced at the most distant or south abutment pier, receding pier by pier to the north bank; and by this ascending by time to time in their wok, they felt no more apprehension of danger when on the highest, than at first on the lowest gangways.

He was proud of his safety record, but did note rather callously that one man had fallen to his death, but that it was due to 'carelessness on his part'.

The aqueduct is one of the great pioneering engineering triumphs of the age, but it had one little homely touch. The joints between the iron plates were made watertight by seals of Welsh flannel that had been soaked in white lead.

The opening of the aqueduct on 26 November 1805 was a very grand affair, with speeches, marches, bands playing and guns firing a salute. Banners with messages were carried, some of which were quite fanciful:

> Here conquer'd Nature owns Britannia's sway
> While Oceans' realms her matchless deeds display.

Others were more prosaic, but perhaps rather more to the point:

> Success to the iron trade of Great Britain, of which Pontcysyllte Aqueduct is a specimen.

There has been a certain amount of controversy in recent years about who should get the credit for Pontcysyllte. Traditionally Telford was given the laurels, but the late Charles Hadfield, the doyen of canal historians, argued very forcibly that he did not deserve the praise. William Jessop was the chief engineer and Telford only his assistant. It is certainly usual to think of the chief engineer as responsible for the works on a canal, and the plans would have needed his approval, but Telford himself wrote, in a letter about aqueducts dated 18 March 1795, 'I have just recommended an Iron Aqueduct for the most considerable, it is approved, and will be executed under my direction, upon a principle entirely new, and which I am endeavouring to establish with regard to the use of Iron.'

It seems quite clear that the idea was Telford's, but the final decision would have been Jessop's, and we simply do not know to what extent both men were involved in design details. But Telford was quite right in that he was doing something very important for the iron industry, in extending its use in major construction projects.

Telford would go on to build other iron aqueducts, notably on the Birmingham & Liverpool Junction Canal, now known as the main line of the Shropshire Union. Later in his life, when he was already established as the country's leading engineer, he was asked to improve on what had become one of the busiest canals in the country, the Birmingham, originally designed by James Brindley as a typical early canal that followed the natural contours of the land, so that it snaked round every obstacle in a series of extravagant curves.

Telford was asked in 1825 to suggest how it could be improved. He planned for a straight route that would slice through the low hills in deep cuttings and cut right through the wandering line of the old canal. This involved some major engineering works, where the new canal was in a deep cutting when it sliced through the old. At one point it cut off the arm that had been built to reach a steam pumping engine, an essential part of the

The great Pontcysyllte Canal Aqueduct.

Repairs under way on a cast iron aqueduct on the Shropshire Union Canal.

water supply system. Telford decided the best solution was to carry the old arm over his new canal on an iron trough aqueduct. He had always rather fancied himself as an architect and now he had such authority that he could indulge his whims.

The Engine Arm Aqueduct is essentially a simple iron trough like many others, but he decided to embellish it. When he began his career classicism was in vogue, but now the taste had shifted to the more elaborate Gothic. He decorated the ribs of the arch carrying the towpath with a series of pointed arches. It was a foretaste of things to come, as the decorative use of iron was developed for many different structures and buildings.

If Telford had done no more than introduce the concept of the use of iron in canal structures, it would still have been an important contribution to the industry, but he did a great deal more. When he took on the engineering job on the Ellesmere Canal he did not give up his post in Shropshire. The floods of 1795 that had swept away Clowes' work at Longdon-on-Tern also carried away the old bridge over the Severn at Buildwas. As county surveyor, Telford was called on to design a replacement. He would obviously have known of the original iron bridge over the river, but he was sure he could

Telford's first iron bridge, crossing the Severn at Buildwas. (Shropshire Record Office)

improve on the design. One of the problems created by the high-arched original bridge was a tendency for the abutments to be forced outwards, compressing the arch. He realised that a flatter arch could be used to overcome the problem, and he worked with the Coalbrookdale Company to see how this could be accomplished.

The magistrates approved the new design, noting that at an estimated cost of £3,700, the price was 'considerably lower than if the same had been erected of stone'. This was an elegant design and it is notable that, although it was 30ft longer than its famous predecessor, it used far less iron – 173 tons against 378 tons. Telford had thought specifically of how to build in iron, without giving too much thought to earlier building techniques. Although he had begun his working life as a mason, he felt that Darby and his colleagues had not 'disentangled their minds from the form of a stone arch'.

The bridge no longer exists, but it did survive long enough to be photographed. It was built on what Telford described as the 'Schauffhausen principle'. The outer ribs spring from the foot of the abutments right to the top of the railings in a graceful curve that rises just 24ft in a 130ft span.

Telford was to go on to build a number of splendid iron bridges, including two more across the Severn. One of his most elegant structures can still be seen in Scotland at Craigellachie, where it crosses the River Spey. It carries a plaque recording the fact that the ironwork was cast at Plas Kynaston, which suggests that the parts must have started their journey by being loaded into a boat that would have been towed across the Pontcysyllte Aqueduct. He also returned to building in iron on the Birmingham Canal, where a deep cutting was crossed by the magnificent Galton Bridge.

But the Birmingham Canal is also noted for another way in which iron could be used. The canal was built to a uniform width, which means that the bridges all had the same span. This, in turn, meant that bridges could also be standardised, and that made them ideal for iron construction. The same patterns could be used over and over again to cast the parts for many bridges. Telford presented the idea to a new company, the Horseley Ironworks at Tipton, who had built their first furnaces in 1809–1810. They soon showed themselves to be efficient and innovative. Their bridges were like the familiar flat packs of today: they came with standardised parts that could be assembled on site. There were two flat arches, each with the handrail as an integral part of the casting. These were joined together by a central locking plate. The bridge was then completed by a footway consisting of iron plates bolted together. Horseley bridges can be seen on a number of canals in the Midlands.

A Telford bridge crossing a Telford canal – Galton Bridge, Birmingham Canal.

Telford was always prepared to look at new construction methods. In 1814 he was faced with a challenge: to construct a bridge across the Mersey at Runcorn. When he visited the site he found that, although the banks were solid, there was no firm footing on the riverbed to support arches and it was too wide to be crossed in a single conventional arch. At this point he began to think about using a different technique: he would build a suspension bridge, with a deck suspended from wrought iron cables.

Before any work could begin he needed to satisfy himself that the cables would be strong enough. There was only one place to turn to for testing. Lieutenant (later Captain) Samuel Brown had developed and patented wrought-iron chains for use in the navy to replace the old rope cables. He set up in business with a cousin, Samuel Lennox, to form Brown Lennox Chain Makers of Pontypridd. The company was a great success – chains made by the company can be seen in the background of the famous portrait of Brunel beside his ship, the *Great Eastern*.

Brown set up a chain-testing establishment in London, and it was here that Telford came to try out his ideas. Some 200 experiments were carried out, at the end of which he had worked out that a 1in square rod of iron could suspend a load of 27 tons. On this basis, he decided that a 1,000ft suspension bridge would be practical, using cables made of thirty-six ½in square iron cylinders joined by buckles at 5ft intervals.

In the event, the bridge was not built but the idea of a suspension bridge and the research were not wasted. Brown designed a suspension bridge to cross the Tweed, the Union Bridge completed in 1820, which remains the oldest suspension bridge in Britain, though James Finley had been using the same technique in America since 1800. Nothing, however, had ever been attempted to cope with the problem that faced Telford on one of his most important assignments.

In 1817, Telford was appointed as chief engineer to create a modern road that would carry the Irish mails to and from the port of Holyhead. There were many challenges to overcome, among them the need for two important crossings, one over the Dee at Conwy and the other, a far more daunting challenge, to cross the Menai Straits that separate the mainland from Anglesey. The specific problem with the latter was the need to have a bridge that would allow the tallest-masted ships to pass underneath. He decided that the only possible solution was to use a suspension bridge, and extended the idea to use similar systems for both crossings.

The Conwy crossing was the more straightforward, but the straits presented major problems. The first was the necessity to keep the seaway through the straits open at all times, but it was a notoriously rough passage. The necessary materials for the construction had to be brought by boat, and during heavy storms in 1820 the supply schooner was all but wrecked and two other smaller vessels severely damaged. We are fortunate that the resident engineer, William Provis, kept a record of the whole construction process, which was fraught with difficulties.

It began with the construction of the two towers and approach roads. The rocks had to be levelled at the Anglesey end of the bridge to provide a platform for the southern tower. The suspension bridge would have to be 579ft long to span the gap and would rise 100ft above the water.

The towers were massive undertakings, but here Telford was working with materials with which he was well familiar. When it came to dealing with ironwork for the bridge he was entering unknown territory and was understandably cautious. He set up a system for testing 9ft-long links and allowed a 100 per cent safety margin, and once they had passed the test they were heated and dipped into oil to prevent corrosion. He was also unsure what strains would occur when the links were joined, loaded and stretched under load. As no one seemed able to supply the information, he set up his own test rig. A length of chain was taken to a nearby valley, and then attached to a solid support and fastened to a crank at the other end.

This was held in shear legs and different loads were tried by simply adding more weights.

The next issue to be decided was how long the rods that would connect the curved chain to the horizontal deck needed to be. This would be a very basic problem for a modern engineer, who would work out the calculations on paper. Indeed, in the book that was written to accompany the opening of the bridge, there is an appendix at the back setting out the appropriate mathematics. But these were calculations made after it was completed, not before work began.

Telford took a pragmatic approach, as Provis explained: 'It is true that ordinates might have been determined by calculation, but with a practical man an experiment is always more simple and satisfactory than theoretical deductions.' So a scale model was built to determine the distances between different points on the curved chain and the deck. He did, however, receive some expert advice from a notable scientist, Gilbert Davies – the man who had helped Trevithick design his locomotive. He suggested that the towers should be higher than originally planned, to allow a steeper curve to the chains which, in Davies' own words, 'would double the strength at the expense of 15 Thousand Pounds'.

Telford realised that his original plan of simply securing the chains to the tops of the two towers would create forces that would tend to drag the towers inwards. Instead, he had the chains run over rollers on the towers to be anchored at ground level. At the Anglesey end this involved cutting tunnels deep into the rock, but when it rained they filled with water, risking corrosion of the ironwork. It was decided to create a drainage adit to carry off the floodwater. However, this had to be cut through solid rock and there was only space for two men to work at the rock face at any one time. There were two shifts, and four men were employed on the work, hacking away at the rock for fourteen months.

One of the critical days for the whole project was 26 April 1825: the first chain was to be set in place. A huge raft, 400ft by 6ft, was constructed and manoeuvred into position with the help of men working from platforms built out from the banks. The chains were assembled in two sections and the ends dropped down from the towers to be joined together on the raft. When everything was in place, the chains were slowly raised by capstans turned by a workforce of 150 labourers.

To help ensure that the two teams worked in complete co-ordination a band was employed to play in a strict tempo to which they could keep time.

At first, the men galloped round the capstan at a good speed as the slack was taken up on the block and tackle. Then, as the weight of the chains began to tell, the pace got steadily slower. In an hour and a half the job was done. Telford made a final check and declared everything was in order. It was a triumph, celebrated by two of the more daring workmen who walked across the Straits on the chains.

Everything seemed to be going really well. The rest of the chains were set in place and the deck suspended. It has to be remembered that this was a pioneering work: nothing on this scale, let alone using this technique, had ever been attempted before, and it was inevitable that some unforeseen problem would arise. It did. No one had allowed for what might happen in a high wind. It had already been seen that before the deck had been attached, the chains were swinging in high winds, but once the rigid deck was attached it was assumed that the motion would be damped down so far as to be scarcely noticeable.

It was opened on 30 January 1826 and the first mail coach crossed and all seemed well. But gale force winds still caused unacceptable vibrations and twisting of the deck, and some of the vertical rods were broken. Further strengthening was needed, but it seemed it had not been enough. One of the mail coaches toppled over on a later date and it was claimed that the movement of the bridge was to blame. A careful investigation was made, and the true cause was unearthed: the coachman was blind drunk. Telford was vindicated and his bridge still stands today.

The other suspension bridge, at Conwy, was completed with very little trouble, and the vital link to Holyhead was open. The work on constructing the bridge at the Menai Straits has been described in some detail because it was such a pioneering work and so important for future developments. The engineers who were moving from traditional construction materials – stone, brick and timber – had to solve problems for which there were no precedents.

The credit for developing suspension bridges should be shared between Brown and Telford. Brown had also built his suspension bridge, the Union Bridge, across the Tweed and it was his early work on chains that made the appropriate technology available. Telford took the idea, showed how otherwise impossible obstacles could be overcome and opened up a whole new chapter in the history of both civil engineering and the use of iron.

The most famous successor to the Menai Bridge must be the Clifton Suspension Bridge. Much has been made of what later commentators have described as 'Telford's ludicrous Gothic design', which had two immense towers rising up from river level all the way to the top of the gorge to carry the chains. In contrast to that, the young Isambard Kingdom Brunel produced a far more elegant solution, with towers at the top of the gorge – but this had the benefit of hindsight.

Telford had first-hand experience of the difficulty he had encountered in stabilising the Menai Bridge in high winds, and had decided that to build anything longer would be to court disaster. But to span the gorge at the higher level would require a span of over 700ft. Perhaps he was being overcautious, but his caution was based on experience. Brunel had the boldness of youth on his side and has emerged as the hero of this story. His fame has overshadowed that of Telford, but it is worth pointing out that the Clifton Bridge was not actually completed until after Brunel's death.

Telford lived long enough to see the start of the next transport revolution, the arrival of the steam railway; an innovation which, as a convinced canal man, he regarded with deep suspicion. Brunel, by contrast, was to become one of the leading figures in this revolutionary movement. This new age was to make even greater demands on the iron industry.

6

STEAM AND IRON RAILS

The spread of railways throughout Britain in the nineteenth century, a movement that was soon carried around the world, made huge demands on the ironworks. It is almost impossible to estimate just how much iron was consumed in manufacturing rails alone, but we can come up with some very crude estimates.

A survey taken at the beginning of the twentieth century showed that the main companies had roughly 40,000 miles of track between them, but that did not take into account the extra rails for sidings and so forth, not to mention a considerable number of lesser companies. Add to that the quantity of rails exported from Britain to be laid abroad in the early years of development and one can safely say that not less than 50,000 miles were laid, and as each track consists of two rails, that gives a figure of 100,000 miles of rail. What does that represent in terms of weight or iron needed? Again, there can be no firm answer as different companies used different weights of rail, but the final figure could be not less than 6 million tons. That is a colossal quantity of iron, and does not include all the other uses for iron that the railways called forth, from the chairs in which the rails sat to the engines and rolling stock that ran over the new track. All this happened in less than 100 years.

The last time railways got a mention, Richard Trevithick had just demonstrated his steam locomotive on a tramway in South Wales. The engine had cracked the brittle cast-iron rails, and when he sent another engine up to a colliery in the north-east of England, the same thing happened again. He made one last attempt to interest the world at large in his invention by running a little passenger train round a circular track near the site of the present Euston Station, but no backers came forward. Shortly afterwards, he was invited to supply pumping engines for a silver mine in Peru, and when there was trouble installing them he went over himself to sort out the problem. It was supposed to be a quick visit: in the event he was away for eleven years, and in his absence the system he had tried so hard to promote suddenly burst into life.

The Napoleonic Wars provided the impetus for change. The price of fodder was rising, and owners of coal mines found that the cost of feeding the horses that still hauled the wagons on the tramways linking the collieries to navigable rivers and canals was becoming exorbitant. A steam locomotive seemed the obvious answer: it needed to be fed with coal, but that was the one commodity the colliery had in plenty. But, the troublesome problem of broken rails still had to be solved. It seemed intractable. As it was the heavy engine that caused the damage, the obvious solution was to use a lighter locomotive, but a lighter engine was not powerful enough to do the job.

One man who set out to find a solution was the manager of the Middleton Colliery near Leeds, John Blenkinsop. He consulted with a local engineer, Matthew Murray. Murray was a steam enthusiast and had already crossed swords with James Watt, who, as usual, was fighting to maintain his own monopoly. Blenkinsop and Murray realised that they needed to increase the tractive power of a light engine, and they came up with an ingenious solution.

Instead of simply having the locomotive running on smooth rails and relying on friction to keep it from slipping, they added a third, toothed rail. This engaged with cogs turned by the locomotive's engine: the rack and pinion system that is now familiar from mountain railways. Ideally, they would have liked two rack rails, one to each side, to keep the system in balance or, failing that, a single rail set in the centre of the track. The first was considered too expensive and the second impractical as horses were still using the tramway and the gap between the rails had to be kept clear. So the rack had to be set on just one side, which produced an awkward twisting effect. However, it worked. In June 1812 the first train set off along the line, and soon two locomotives were in regular service. Each was able to haul 85 tons – a load that was later increased – with no damage to the rails.

Blenkinsop was a great propagandist for his system. His first locomotive, named *Salamanca*, appealed to the patriotic spirit of the times, commemorating Wellington's most important victory in the Peninsular War, and *Prince*

Regent was the second (no one ever lost out by flattering royalty). He busily encouraged others to use his system. He sent a detailed estimate to the manager of Oxledge Colliery, spelling out the savings to be made by using his locomotives instead of horses.

He gave the annual cost for running a 5½-mile tramway at £9,653 13*s*, of which the biggest item was fodder for eighty-one horses at £50 each – men came cheaper at a mere £40 a year. He knocked £200 off the horse feed bill for selling the manure for agriculture. The cost for rack rails was £6,247, but £4,465 could be covered by selling most of the horses. The annual running cost came down to a very encouraging £1,468 4*s* a year, less than a quarter of that for the old tramway system.

This novel form of transport attracted a lot of attention from all kinds of people. Grand Duke Nicholas of Russia came for a look, and was so impressed that when he became Tsar he made sure his country was among the first to build a steam railway. The owners of Killingworth Colliery near Newcastle-upon-Tyne sent their young engine man along. It is doubtful if he received as much attention as the Russian nobleman, but he was to play a far more important role in railway history. His name was George Stephenson.

It must have been obvious to engineers at the time that there was a future for steam railways, but that it probably would not lie with a cumbersome rack and pinion system. The answer had to be in producing better rails. Stephenson, encouraged by his colliery employers, had designed his own first locomotive, the *Blucher*, for the Killingworth Colliery in 1814, and he had already found an improvement. Earlier tramways, such as that on which Trevithick had made his experimental runs, had L-shaped rails and the locomotive was fitted with plain wheels.

Killingworth was different: the tramway there was fitted with edge rails to take flanged wheels. This type of rail was becoming increasingly popular. William Jessop, the great canal engineer, was a partner with another engineer, Benjamin Outram, in the Butterley Ironworks in Derbyshire. There, he developed the first rail that was I-shaped in cross-section and this was laid on a tramway in Leicestershire. A common form of edge rail was further strengthened by having the underside curved, so that it was thicker in the middle than at the ends. The shape gave them the popular name 'fish-bellied'.

There was still a recurrent problem with the tramway and plateway systems. To keep the centre clear for the horses, the rails were usually set on stone blocks. A hole was drilled in the middle of the block and filled with a wooden peg, into which an iron spike was driven to hold the rail. The problem was caused by the block shifting and twisting the rail. George Stephenson was encouraged by his employers to work closely with the Walker Ironworks in Newcastle, and he was seconded to them for two days a week at a salary of £100 a year. There he began working with the senior partner, William Losh, to devise a new type of rail and chair. Up to then, rails had simply met end to end, but the Losh–Stephenson rail had a lap joint, one rail overlapping the other. They were designed to sit in a curved chair, so that even if the block moved, the rails could remain steady. They had taken cast-iron rail manufacture as far as it could go, but the future lay elsewhere.

By the end of the eighteenth century an alternative was starting to appear. Thanks to Henry Cort's invention of the rolling mill it was now possible to produce wrought-iron rails as well as cast iron. When Stephenson moved on from providing locomotives for local colliery lines to promoting the first public railway to be specifically designed for steam locomotives (the Stockton & Darlington), his first thought must have been to lay the rails that he had developed with Losh. When the line had first been proposed, one of the engineers consulted had been Robert Stevenson of Edinburgh, one of a family that would become famous for lighthouse building and which included the novelist Robert Louis Stevenson.

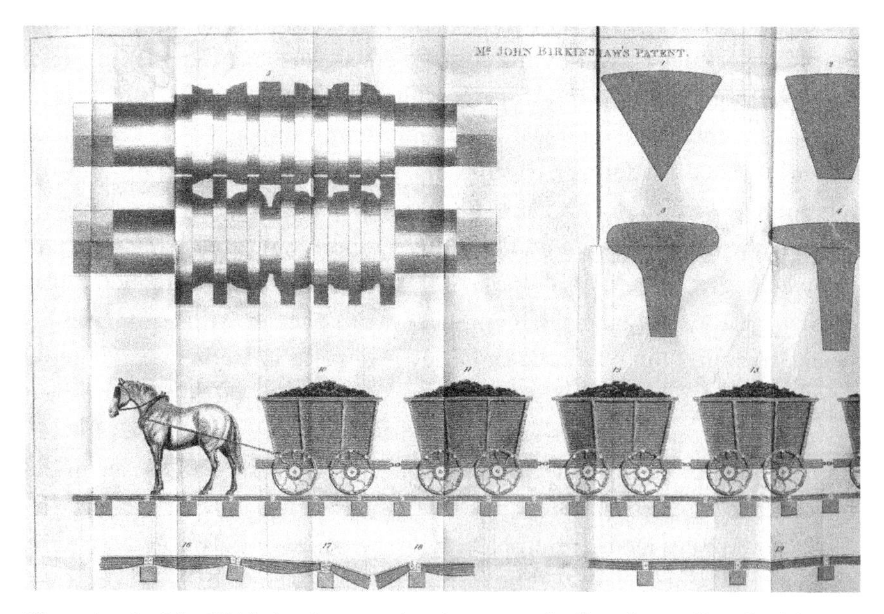

Illustration for John Birkinshaw's patent, showing grooved rollers, the profiles of rail that could be produced and a typical horse-drawn tramway.

Stephenson had heard about a new type of wrought-iron rail with an 'I' section being produced in 15ft lengths, later extended to 20ft, that had been developed by John Birkinshaw. He wrote to Stevenson on 28 June 1821. There was still talk at this time of linking the two towns by a canal instead of a railway and Stephenson was keen to point out that both rails and locomotives were being improved:

> With this you will receive three copies of a specification of a patent by John Birkinshaw of Bedlington, near Morpeth.
>
> The hints were got from your Report on Railways, which you were so kind as to send me by favour of Mr Cookson some time ago. Your reference to Tindal Fell Railway led the inventor to make some experiments on malleable iron bars, the result of which convinced him of the superiority of the malleable over the cast iron – so much so, that he took out a patent.
>
> These rails are so much liked in this neighbourhood that I think in a short time they will do away with the cast-iron railways.
>
> They make a fine line for our engines, as there are so few joints compared with others ... I am confident a railway on which my engines can work is far superior to a canal. On a long and favourable railway I would stent my engines to travel 60 miles per day with from 40 to 60 tons of goods.

The Birkinshaw rails had several advantages, notably the material used and greater length, and it is to Stephenson's credit that he recognised their values and recommended using them, even though he would have made a substantial profit if he had stuck with the Stephenson–Losh rails. As it was, it ended his association with Losh and the Walker Ironworks. He was able to point out that, though the wrought-iron rails cost £12 10*s* per ton compared with £6 15*s* per ton for cast iron, their weight of 28lb per yard was about half that of cast iron, so the actual prices were much the same.

The Stockton & Darlington Railway was a curious mixture of the old and the new. It still used stone sleeper blocks, simply because, although the freight was all moved by steam locomotives, passengers on the line had to make do with a horse-drawn stagecoach fitted with flanged wheels. Where the line met hilly country, the wagons were hauled up the slopes by means of cables and stationary engines, simply because no one was yet convinced that locomotives could do the job. And, in spite of Stephenson's enthusiasm for the new Birkinshaw rails, part of the track was still laid with cast-iron rails.

It was clear that Losh was deeply offended by Stephenson changing to the new wrought iron system, as Walker's did not even supply these sections; instead they came from the Neath Abbey Ironworks in South Wales. The line was a cautious step towards a railway future – a stage that would be completed by Stephenson's next major project.

In spite of the success of the Stockton & Darlington, the directors of the proposed line to link Manchester and Liverpool were unconvinced that the future lay with locomotives. There was a considerable body of opinion that a system of stationary engines hauling rolling stock along from one to the other by cables was a better solution. It was decided to settle the issue with a test designed to see if a locomotive could do the job. Engineers were offered a premium of £500 above the cost price for a locomotive which 'shall be a decided improvement on any hitherto constructed', and they set out specific requirements.

The engine had to 'consume its own smoke', which, in effect, meant that it had to be fired with the smokeless fuel coke instead of coal. It should have two safety valves, one of which had to be out of the reach of the driver – this was to prevent the pernicious habit of fastening the valve down to get higher pressure. This practice had cost more than one engine man his life when the boiler exploded. The most important requirements were the sizes of the engines and the performance expected of them. Six-wheeled engines could weigh up to 6 tons, four-wheeled up to 4½ tons. The performance required of the larger engine was that it:

> … must be capable of drawing after it, day by day, on a well-constructed Railway, on a level plane, a Train Of Carriages of the gross weight of Twenty Tons, including the Tender and Water Tank, at the rate of Ten Miles per Hour, with a pressure of steam in the boiler not exceeding Fifty Pounds on the square inch.

George Stephenson was anxious to enter the contest, but he was fully occupied with the demanding work of building the line and he was desperate to get the help of his son, Robert. But there appears to have been some family quarrel and Robert had taken work as a mining engineer in South America. Now the message went out – he was needed at home.

What happened next is so remarkable that, although basically it has nothing to do with the story of iron, it is so unlikely that no one could ever have dreamed it up and it is worth the telling. Robert made his way to the port of Cartagena to embark for Britain. While he was waiting for the ship, he heard there was another Englishman in the town who had suffered a terrible journey in trying to find a route from the gold mines, in which he had developed an interest, to the sea.

En route he had lost most of his possessions, had a boating accident in which he was nearly devoured by an alligator and was now sick and penniless. He was Richard Trevithick. Robert lent the railway pioneer enough money to get home, while he himself set sail to revolutionise the new industry that the older engineer had started. One would dearly love to have a record of their conversation, for surely they must have talked about locomotives and railways, and young Robert might even have got some useful hints from the older man.

There were three steam locomotives entered for the public competition, which was to be held on the line at Rainhill in October 1829. One was a development of the type of locomotive already in use, designed by Timothy Hackworth, who was in charge of locomotives for the Stockton & Darlington. The second was a lightweight affair with a vertical boiler designed by John Braithwaite and a young Swedish engineer, John Ericsson. This latter looked like a real speedster and was the crowd's favourite. George Stephenson took a more practical view of this sporty model: no guts, was his verdict, and events proved he was right. The third locomotive was the Robert Stephenson engine, *Rocket*, and as everyone knows, it won the day. It not only easily satisfied the rules of taking its load the whole distance from Manchester to Liverpool and back at over 10mph, it also proved itself capable of far greater speeds, reaching an unprecedented 30mph.

It was just as well that *Rocket* was the winner because it contained all the elements on which the future of locomotive design would rest. It had a multi-tubular boiler, which ensured that steam could be raised efficiently. Instead of vertical cylinders it had cylinders that were closer to the horizontal, and the exhaust from the cylinders was sent up the chimney, increasing the draught to the firebox. The latter idea had already been used by Trevithick, but forgotten by later engineers. (Was this one of the things they perhaps discussed in Cartagena?)

The steam locomotive had won the day. Stationary engines would still be used for a time to haul trains up steep slopes, but horses would be limited to pulling carriages and wagons on roads, not on rails. It was the start of a rush of railway construction, not just in Britain but in many countries overseas. It also presented new challenges for the ironworks and ironmasters.

The earliest locomotives had been slow and comparatively crude, with quite wide tolerances. Now, with new designs and increased speed, accuracy was essential. Following the change in position of the cylinders in *Rocket*, later engines had cylinders that were more or less horizontal. This meant that the fit between piston and cylinder had to be very precise to prevent uneven wear. Soon, the companies were establishing their own works to build and repair locomotives and rolling stock. The Stockton & Darlington Works were established at Shildon, and whole new towns grew up around works, such as those of the Great Western at Swindon. And because Britain was the pioneer in the new transport system, other countries were soon looking to this country to supply their needs for everything from hardware to expertise.

Grand Duke Nicholas was now Tsar Nicholas I, and the enthusiasm he had shown on his visit to Middleton Colliery had not abated. There was an abortive attempt by Russian engineers to build a line and its locomotives, but the first successful line, which ran from Tsarokoe Selo, home of the Imperial Summer Palace, to St Petersburg, relied heavily on British expertise and materials. Russia did have its own iron industry but it proved cheaper to order rails from Merthyr Tydfil than to carry them overland from the Urals. Rolling stock was ordered, and four locomotives, two from the Robert Stephenson works in Newcastle, first established to build *Rocket*, the other two from Timothy Hackworth's works at Shildon. It was not just a question of sending material over to Russia; British engineers had to go to explain what everything was and how it worked.

Timothy Hackworth's 17-year-old son, John, went to Russia with the Shildon foreman and a team of fitters. He sent back a nonchalant account of his adventures that sounded as if they might have come from a boy's comic book. They took a flask of spirits to keep themselves warm for the sleigh ride from the port, but it froze solid and, at one point, their sledge was chased by a pack of wolves. When the team arrived they found a tower of Babel – trying to give instructions to the locals, while the conversation constantly changed from English to Russian, Flemish or German.

Meanwhile, back in Britain the rail network was spreading rapidly, with more major cities being linked together. Engineers were still working out the best form of rail to use. Joseph Locke was working on the Grand Junction Railway from Birmingham to Warrington, on which construction began in 1833. Along the way it passed Crewe Hall, near what was to become one of the biggest railway workshops in the country.

Now that horses were no longer being considered, there was no need to leave a gap between the rails and the engineer was able to use wooden sleepers. He also used a different type of rail, in cross-section looking like a dumb-bell. The idea was that as one end got worn it could simply be turned over. That did not work in practice, so the design was adapted to be more like a figure of eight, with the upper part larger than the lower. This bullhead rail was to become the standard form, but in those early days there were many alternatives to choose from.

One system was entirely different from the rest, the one used by Isambard Kingdom Brunel for his Great Western Railway between Bristol and London. Brunel was always the odd one out, rarely taking, or even listening to, advice. For a start he derided Stephenson's paltry gauge of 4ft 8ins – an extra ½in crept in later. He went for a broad gauge, with rails set 7ft apart. He developed his own unique system for laying the track. Instead of sleepers across the track, he laid down longitudinal baulks of pine, topped with hardwood, on which the rails were set. The baulks were then braced by cross ties, which ran right across both sets of tracks. These ties were spiked into piles driven deep into the ground. The idea was to provide an absolutely rigid system, with no give in it at all.

He decided the Birkinshaw rails were not for him either, and he used a rail that was basically like an inverted letter 'U' with wide serifs or a broad-brimmed hat. The author has a length of this rail as a doorstop in his study – it measures 12in from side to side and rises to a height of 4½in in the centre. The rails weighed 55lb a yard, making them twice as heavy as the Birkinshaw rails – and in later years they were to be even heavier.

Rolling rails was a demanding job. The work started with a billet of wrought iron that might weigh around 4cwt, heated to a white heat. It then had to be fed through the rollers, which were grooved to create the appropriate shape and were turned either by a waterwheel or steam engine. These great hunks of hot metal had to be handled by a team of men, passing the billet from side to side until it achieved its final form. It was not just the production of rails that kept ironworks busy, and many traditional firms found it worth their while to begin specialising in railway works.

Wortley Top Forge was founded in 1640 in a lovely wooded valley just north of Sheffield. There was an extensive modernisation programme in 1713, when most of the present buildings were constructed. A weir was built across the River Don to divert water to fill the forge pool and provide the power for three wheels. One worked the bellows for the furnace; the other two worked the giant hammers of the forge. Sometime near the end of the eighteenth century they introduced a puddling furnace, and around 1835 they started a specialised business manufacturing wrought-iron railway axles to meet the rapidly growing demand.

It was a complex process. Each axle was made up of sixteen iron bars roughly 1¾–2in square. Four of the bars were 45–51in long; the other fourteen were 33–36in. They were arranged into what was known as a 'faggot', a four by four square, with the longer bars in the centre. One end was flush and the long bars protruded from the other, providing a handle for manipulating the block. The whole faggot was held in place with an iron hoop.

The work involved a five-man team. One was in charge of the furnace that heated the flush end of the bar to a white heat. The crane operator then moved the hot bar and placed it under the hammer. Two men manipulated the forging and shaping of the faggot. At the start, sand was sprinkled on the faggot and as the pounding started white-hot flakes of slag flew off in all directions. The men had face masks of wire gauze, long leather aprons, iron-protected boots and shin guards. Even with this protection, burns were a commonplace hazard of the job.

At first the men consolidated the faggot into a solid, welded piece and then increased the rate of the hammer, by allowing the waterwheel to turn faster. This was controlled by the fifth member of the team, usually a boy. At full speed, the hammer was striking at a rate of 175 strokes a minute. The faggot was regularly reheated and turned by the two hammer men until it was perfectly round, after which the protruding end was cut away.

It seems remarkable that, in the age of the steam locomotive, axles were being made by water-powered hammers, but it was cheap and efficient – it has been estimated that one of the hammers had a force that was the equivalent to that of a 25cwt steam hammer. They continued making axles here right up to 1912, when they were turning out 250 a week.

The working conditions at the forge were wretched. The men worked in two shifts, one starting at 6 a.m. going through to 5 p.m.; the next starting at 6 p.m. and ending at 5 a.m. The noise of the hammers was deafening, and the ground shook. The heat was intense and the slag particles still embedded in the foundry wall are a permanent reminder of what it must have been like to work under a hail of flying, white-hot fragments.

Iron was also to play a role in the civil engineering structures necessary for building a railway, with bridges representing the more challenging projects. When the pioneering Stockton & Darlington was extended across the Tees, the engineers faced the same sort of problem that had exercised Telford when he had to construct road bridges over navigable rivers. Again, the Navigation Company insisted that, in order to avoid any obstruction to shipping, there should be no piers built on the river bed. So the company decided to use the solution that had worked on other crossings such as the Menai Straits. They would have a suspension bridge, and they turned to Captain Brown who had pioneered the technique. The result was a disaster.

Robert Stephenson described what happened: 'Immediately on opening the suspension bridge for railway traffic, the undulations into which the roadway was thrown, by the inevitable unequal distribution of the weight of the train upon it, were such as to threaten the instant downfall of the whole structure.' The bridge had scarcely been opened before it was closed to all locomotive-hauled trains. In 1842 Robert Stephenson designed a successful iron girder bridge to do the job.

Stephenson had learned a lesson about crossing navigable rivers, and never considered using suspension bridges. But he still had to find suitable alternatives. He was appointed as engineer for the Chester & Holyhead Railway and needed to build a bridge across the Dee at Chester. He originally planned a conventional five-span brick bridge, but was told that the foundations would be insecure, so opted instead for an iron bridge. He used 87ft-long girders that consisted of I-frame cast-iron members reinforced by wrought-iron trusses.

Originally the bridge was only being used for constructors' trains while work went on along the rest of the line, but then the Shrewsbury & Chester Railway were given permission to use it. On 24 May 1847 a passenger train was crossing the bridge when the driver became aware of a strange noise and felt vibrations under his engine. He opened the regulator wide in an effort to get off the bridge as quickly as possible. The locomotive was clear when the end girder gave way, plunging the rest of the train down to the river. The fireman was unfortunately breaking up coal in the tender and he was killed, along with a guard and two coachmen. One passenger died and sixteen were seriously injured.

There was a public inquiry and Stephenson was able to call on an array of eminent engineers to give evidence that the design was sound and the bridge failure must have been caused by some accident, probably a wheel breaking on the tender as it crossed the bridge. Only the engineer of the Shrewsbury & Chester Railway declared that the fault lay with the fundamental design. In the event, the jury brought in a verdict of accidental death and declared the case against Stephenson was 'not proven'.

Floating the box girder for Stephenson's railway bridge at Conwy, with Telford's suspension bridge in the background. (Elton Collection, Ironbridge Gorge Museum)

If a different decision had been taken, his career might well have been over. As it was, he turned away from cast-iron bridges and now had to find a different solution to the next major headache he faced, the one that had earlier been dealt with by Telford – taking the lines to Holyhead and crossing the Dee at Conwy and the Menai Straits.

With a suspension bridge and a cast-iron bridge both ruled out, Stephenson had to venture into unknown engineering territory. He did have one physical feature he could use – the Britannia Rock lay in the centre of the Straits and could become the foundation for a supporting pier. That still left an immense span of almost 500ft from the rock to the shoreline in each direction.

He began a series of experiments with wrought-iron girders, and in order to achieve the essential strength and rigidity he came up with the novel idea of using a hollow box girder with the rails actually running inside the box. Nothing like this had even been considered before, and Stephenson spent several sleepless night worrying about the tubes. He wrote to a colleague, Thomas Gooch, after the event:

> Often at night I would lie tossing about seeking sleep in vain. The tubes filled my head. I went to bed with them and got up with them. In the grey of the morning when I looked across the square it seemed an immense distance across to the houses on the opposite side. It was merely the same length as the span of my bridge!

Fortunately, a new technique for rolling wrought-iron plates had been developed in 1830 specifically for making locomotive boilers. He built a one-sixth scale model to test for strength, and also called upon Professor Eaton Hodgkinson, a structural engineer who, unusually for that time, had used mathematics as well as experiment to determine the strength of structures. He was in favour of Stephenson's idea, but with reservations: 'If it be determined to erect a bridge of tubes, I would beg to recommend that suspension chains be employed as an auxiliary, otherwise great thickness of metal would be required to produce adequate stiffness and strength.' Stephenson was unconvinced, but the piers were extended upward to take chains anyway – they were never needed.

As with Telford's suspension bridges, Stephenson decided to start with the shorter Conwy crossing. The plates were riveted together *in situ* to form the massive tube, and when completed it was to be floated into position and raised by hydraulic jacks. On 6 March 1847 the tube was floated out on the high tide. A full account of construction was given by Edwin Clark, who described the events of the day:

> The tube being lifted by the pontoons, began to move off, snapping the small ropes that kept it back; it glided quietly and majestically across the water in about twenty minutes ... At eleven o'clock the deep and rapid Conway was an impassable gulf, and in less than half an hour it was spanned by an iron bridge.

Stephenson now felt able to continue to face the more demanding task at the Straits, where the Britannia Bridge would be built on a far grander scale, with a total span of 1,800ft and standing 100ft above the water. The Britannia Bridge was a success, and the portals were very properly inscribed with the name of the engineer. Further embellishment was provided by the sculptor John Thomas, who carved lions to guard the entrance. They were not universally admired. One contemporary described them as 'of the antique, pimple-faced, knocker-nosed Egyptian kind'.

Stephenson's box girders were to be used again, most dramatically for the railway crossing the St Lawrence River at Montreal. Its size dwarfed even the Britannia Bridge, with a total span of 6,512ft, but fortunately the river was comparatively shallow and ran over a bed of solid rock, well able to support piers. The nineteenth century was a great time for producing statistics for construction programmes, and figures for the bridge are certainly impressive. The twenty-five spans varied in length from 242ft to 330ft, used 9,044 tons of iron and were held together with 1,540,000 rivets. The work force of 3,040 men was helped by six steamboats, 75 barges, 144 horses and two steam locomotives.

Conditions for those working on the great bridge were atrocious. The men who went over from Britain were unprepared for the winter and many suffered from frostbite. Summer brought no reprieve, when cholera swept through the workings. Caissons, temporary dams, had to be built in order to construct the masonry piers, but had to be dismantled every winter, to prevent them being swept away by ice floes thundering down the river. The iron was made in Birkenhead and drilled there, ready for riveting and assembly. Work began in 1854 and was completed five years later, and was considered a triumph, but many questioned whether it made any sense to bring British expertise and materials across the Atlantic and to use technologies devised for very different circumstances.

There were many other important iron railway bridges built in Britain during the nineteenth century, but this type of construction was brought to an end by a catastrophe. In 1879, the viaduct across the River Tay at Dundee collapsed with the loss of ninety lives. The engineer, Thomas Bouch, received the blame for cutting costs and not supervising the contractors with the care he should have done. A year after the accident, he died a broken man. Plans for a similar structure over the Firth of Forth were at once shelved, and when new plans were drawn up they were no longer for an iron bridge: this one would be built of steel. That is the subject of a later chapter, but there was an element in the story of Stephenson and the tubular bridges that will take us to the next stage of the story.

The other engineer to whom Stephenson had turned to for advice was William Fairbairn. He was establishing a reputation for building iron ships, and his yard was one of the few places that contained machinery that could be used to test the strength of iron beams and girders. He worked closely with Stephenson and invited him to see an accidental demonstration of the rigidity of a box girder-like structure.

Brunel's railway bridge across the Tamar, photographed before the present road bridge was constructed, hence the car ferry.

During the launch of the steamship *Prince of Wales* into the Thames at Blackwall in 1847, things had gone badly wrong and the vessel was stuck with 110ft of the hull completely unsupported. In its basic structure, the unsupported section was no different from the proposed iron tube for the bridge, and it withstood the test without damage. It was a great encouragement to Stephenson; it was also a demonstration of the value of the iron ship. A conventional wooden vessel would have broken its back.

Fairbairn was not the only engineer to get involved with Stephenson's pioneering bridges. Brunel, although in many ways a rival of the Stephensons and with very different ideas on railway construction, was nevertheless a good friend of Robert's and made the journey up to Wales to offer his support. He too used a metal tube in bridge construction, but in a quite different way. For the crossing of the River Tamar, he used curved, hollow, elliptical iron tubes from which the bridge platform was suspended: a bowstring bridge. It was to be his last work: he was taken across the finished bridge on a special invalid carriage for the opening in 1859. A few weeks later he was dead. But before that he had made what is arguably his greatest achievement. He had developed an iron liner to cross an ocean.

7

THE IRON SHIP

The use of iron for ships' hulls was intimately connected with the development of steam power. Surprisingly, given British dominance in the manufacture of steam engines, the first successful experiment took place in France, and the innovator was no hard-handed mechanic but the aristocratic Marquis Jouffroy d'Abbans. His steamer, *Pyroscaphe*, made its maiden voyage on the Saône in 1783.

The 1780s were not, however, a good time for scientific and technological revolutions in France, nor indeed for aristocrats. In Britain, there was a first tentative attempt to introduce steam power when, in 1788, William Symington put an engine in a small boat and made several trips on a lake near Dumfries. Among the first passengers was the poet Robert Burns who, sadly, although moved by steam was not moved to write verse on the experience. Encouraged by his success, Symington went on to build a steam tug for the Forth & Clyde Canal. It was an appropriate setting as, unlike most of the English canals, it was a broad waterway able to take vessels of 19ft beam. It also had an unlikely steam connection: James Watt had worked on it for a time as a civil engineer during its construction.

In March 1802 the *Charlotte Dundas* hauled two 70 ton barges for 20 miles up the canal in about six hours, which was quite an achievement for a modest 10hp engine. But the canal owners decided that the churning paddle wheels would damage the banks and the experiment was abandoned. It was an American, Robert Fulton, who had the honour of building the first steamer actually to make it into regular service in 1807, plying between Albany and New York. Britain finally got its first commercial steamer five years later.

Where Symington had become involved with steam power through his work in the mining industry, David Napier's background was in the iron industry. His father had a forge and foundry, based first at Dumbarton then in Glasgow. The young Napier had a conventional schooling, but from the first was more interested in the works than the classroom. There were two steam engines there, one a blowing engine for the furnace, the other used to drive a machine for boring cannon. In Napier's own words, 'I never served a regular apprenticeship but put my hand to everything.' By the time he was 20 he was running the works, and when shortly afterwards his father died, he took total control. He was now able to marry and was to father fifteen children, a bewilderingly large family that he simply referred to as 'the swarm'.

One of the visitors to the works was a hotelier from Helensburgh, who had heard of Fulton's work in America and liked the idea of running steam excursions on the Clyde. He asked Napier to provide the engine and boiler. The engine itself was simple with a single 12½in vertical cylinder, with power transmitted through a crankshaft to the paddle wheels. It had even less power than the *Charlotte Dundas* and in some illustrations it can be seen with a sail rigged to the tall funnel.

This was all new territory for Napier, and he worked more by trial and error than theory:

> I recollect that we had considerable difficulty with the boiler, not having been accustomed to make boilers with internal flues, we made them first of cast iron but finding that would not do we tried our hand with malleable iron and ultimately succeeded with the aid of a liberal supply of horse dung in getting the boiler filled.

The engineers and ironmasters had become used to providing boilers for the early steam engines of the Watt generation; boilers that were little better than oversized kettles. Boilers with internal flues had been introduced by the next generation, originally by men such as Trevithick with his Cornish boiler, and by the locomotive engineers.

Napier was clearly unfamiliar with the changes, and had to set about inventing the technology all over again. His reference to horse dung suggests that his first effort was not altogether satisfactory, and that there were leaks that needed plugging. But he persevered, and the new vessel, *Comet*, was sufficiently successful for him to see a future in steam on the water, even if, as he wryly noted, Bell never got round to paying him.

He described his next step in his own handwritten notes intended for his memoir:

> Seeing steam navigation was likely to succeed I erected new works at Camlachie for the purpose making steam engines where the engines were made for the 'Dumbarton Castle' the first steamer that went up Loch Fyne and for the 'Britannia' the first that went to Campbeltown. Although these vessels did not venture outside the Cumbraes in stormy weather they suggested the idea to a company in Dublin of having steamers between Holyhead and Howth, for which purpose two vessels were built at Greenwell, no expense being spared to ensure success – The engines were made by Mr James Cook at that time the oldest and most respectable engine maker in Glasgow, but when tried on the station the engines were so complicated and cumbersome that they broke down almost every gale of wind, and ultimately were laid up in Kingston Dock near Dublin, as useless, and the idea of making machinery of any kind that would withstand the shock of a heavy sea in a gale of wind was put down an impossibility – Whether it was from pique and not having being employed to make the engines of these vessels or from a conviction that the ocean could be safely navigated by steam I cannot now say, but I commenced I think about the year 1818 to build a steamer on my own account for that purpose called the 'Rob Roy'. I recollect the day before starting on the first trip from Glasgow to Dublin Mr Chas. McIntosh, the celebrated chemist and inventor of water 'proof cloth' saying we should all be drowned.

In the event, the ship survived a battering by storms on the crossing to Ireland and on the return without damage – and without anyone being drowned. Napier was now able to establish a new business with confidence, and was to go on to make his own improvements in engine design. The value of steamers was becoming well established, but the early vessels were still being built with wooden hulls. That, too, was about to change.

Various small iron vessels had been made for the canals, especially the so-called 'tub boats', little more than open-topped iron boxes. One manufacturer, Charles Manby, decided to start making iron steamers at his works at Tipton on the Birmingham canal system. His first vessel, named after his father, was the *Aaron Manby* and it was destined for a working life in France. Although it was a modest vessel, 180ft long by 17ft beam, it was too large to fit into the narrow canals of England, so it was made in sections that were then sent down to the Thames in London for final assembly. In 1822 the paddle steamer made its way across the Channel to Le Havre. Steam and iron had come together for a sea crossing for the very first time.

In a comparatively short time, two important points had been established: paddle steamers could operate in rough seas and iron hulls could also be used at sea. The man who moved the whole industry forward was William Laird, who arrived at Merseyside in 1810 at the age of 30 with the original intention of getting orders for his father's rope works at Greenock on the Clyde. Instead he saw the possibility of developing a new works on the shore across the Mersey from Liverpool. It was to be the Birkenhead Ironworks, though Birkenhead itself scarcely existed.

Laird laid out a new town with all the most modern amenities, from gas lighting to pumped water for every house. By 1829 he had decided to specialise in iron ships, though there were problems in getting them widely accepted. One difficulty was that iron affected that most essential of navigational aids, the ship's compass. For once, the problem was not solved by the practical engineers but by an academic, Professor Airey. He realised that the magnetism from the ship's ironwork was a constant and the compass bearings could be corrected by counter-balancing magnets, placed on either side of the binnacle, the casing holding the compass.

The most important development from this enterprising beginning lies with that wayward genius Isambard Kingdom Brunel. Famously, when public doubt was expressed about the possibility of having a railway as long as one running between Bristol and London, the engineer replied that not only was it possible but he did not see why it should not be extended. Why not, he asked, build a steamer and extend it to New York? Whether he meant this to be taken seriously, or whether it was simply a typical piece of off the cuff Brunel bombast, we'll never know. But it was taken seriously by Thomas Guppy of the Great Western board who enthusiastically backed the idea, and Brunel set about the task of building a transatlantic steam ship.

There were more critics to contend with, notably Dr Dionysius Lardner, author of one of the most popular encyclopaedias of the day. He 'proved' conclusively that the thing was impossible. If a ship could carry enough fuel for the journey there would be no space for cargo or passengers, and if the size of the ship were increased the situation would remain the same; the bigger the ship the more fuel it would need.

Brunel may not have been a professional scientist but he understood physics rather better than the learned doctor. To move a ship through the ocean means overcoming water-resistance – weight is not a problem as it is floating. But water-resistance depends on hull area, which is a measurement given as a squared figure – square feet, for example. Increasing size increases the volume – a cubic measure. In other words, resistance to water does not increase in the same proportion as increase in size. The bigger the ship, the smaller the proportion of the space actually needed for fuel.

Brunel's first vessel, *Great Western*, made its maiden voyage in 1837. It was beaten to New York by the much smaller *Sirius*, which had started earlier but had arrived with her coal bunkers empty, while the *Great Western* still had fuel to spare. Brunel had been right. The future lay with big steamers. But now he faced a problem. There are limits to how big a vessel you can build using timber for the hull before the whole structure becomes fatally weakened.

By now Laird's yard was turning out iron ships, and one of their Channel packets, *Rainbow*, steamed into Bristol. Brunel went on board, and his colleague Captain Christopher Caxton made a number of voyages in order to give his professional opinion on its seaworthiness. Brunel was convinced, not least because iron structures had a double advantage – they were both stronger and could be constructed using less material, as he explained in a report of 1840:

> Suppose all the angle irons or ribs the shelves etc. were all rolled out flat and added to the thickness of the plates forming her sides, when an average thickness of 2ft of timber would be replaced by an average thickness of 2½in of iron, with far better ties, a more compact framework and greater strength, than wood can under any circumstances.

There was no widely accepted method for building an iron hull. This was to be far bigger than any ship ever built before. It was eventually to have an overall length of 522ft, a beam of 50ft 6in at the widest point and a depth to the upper deck of 32ft. If one compares this with one of the finest sailing ships of the nineteenth century, the *Cutty Sark*, the comparable figures are 280ft, 36ft and 21ft. Measured in terms of volume, the *Great Britain* is roughly four times as big.

The technologies used by shipwrights for countless generations were no longer useful. Everything had to be done in quite new and different ways. In a wooden ship the keel was laid down, ribs were set at right angles to it, beams set across between the ribs and the exterior was planked. However, the base of the iron ship consisted of ten iron girders running from stem to stern, above which an iron deck was secured. Five watertight transverse

The author, standing by the propeller of the *Great Britain* during filming for the TV series *Big, Bigger, Biggest*.

bulkheads were built up above that and two longitudinal bulkheads ran the length of the ship. The result was a set of rigid, open-topped boxes. For maximum strength, this would have been topped off by an iron-plated upper deck, but that was perhaps a stage too far away from the traditional, so wooden planks were used.

The main problem faced by the engineers was that the rolling mills could only supply comparatively small wrought-iron plates. The maximum size available was only 6ft long and 2ft 9in wide and about ¾in thick. The difficulty lay in devising a way of fitting them together so as to be completely watertight. The technique adopted was similar to that used for traditional clinker-built wooden boats, in which the strakes, the timbers running fore and aft, overlapped each other. The iron plates were laid in rows, with each strake overlapping the next. They were then riveted together. The vertical joints were secured by iron straps on the inside of the hull.

The whole ship was built in a new shipyard in Bristol and machinery was supplied by Napier. There were shears for cutting plates accurately and punches for cutting the holes for riveting. It seems remarkable to us that holes could be punched through a metal plate using a machine that could be powered by hand, but it was not simply a case of knocking a hole through. The principle was a familiar one from earlier ironworks. A large wheel was turned, which worked through gearing to operate a lever that raised and lowered the punch.

Riveting was arduous work. First the plates had to be carefully positioned and the holes accurately lined up. The rivets generally had domed heads. They were heated, pushed through the punched holes, held in position, and on the opposite side of the plate, the end of the rivet was hammered over until it lay flush with the hull. It has been estimated that a four-man team working a ten-hour day could drive as many as 140 rivets a day. The work was not only very strenuous but the noise was, quite literally, deafening.

In the early stage of the development of the ship, it was still planned, as Brunel's previous vessel had been, as a paddle steamer. There was much discussion over who should supply the engine and what sort of engine it should be. The contract went to Francis Humphreys, a young man who had patented a trunk engine that was being manufactured by Halls of Dartford. One of the fascinating things about the early technology of iron and steam is the way that the same names keep turning up in different contexts. Halls had employed the pioneer Trevithick during the last years of his life – they were clearly a company open to innovators and innovation.

At this stage, the new ship was going to be a paddle steamer, like its predecessor, but with a much larger and more powerful engine. Transmitting the power from the engine to the wheels would require an immense crankshaft, and that was a problem. The existing tilt hammers were all operated as giant levers, and there was a limit to how high the hammerhead could be raised above the anvil. There was not a hammer in the land big enough to forge Humphreys' crankshaft. It was not a problem Humphreys could solve on his own, so he wrote to the manufacturer James Nasmyth for advice.

Nasmyth had enjoyed a remarkably successful career. As a boy he had a private tutor, but spent a lot of time in a foundry owned by the father of one of his friends. He discovered a natural talent for mechanical engineering, and was soon making model steam engines that he sold to help pay for an education at Edinburgh University. But he was well aware that, for a successful career, he would need at least as much practical experience as theoretical knowledge. He managed to gain a place with, arguably, the most inventive manufacturer of machine tools of that, or any other age, Henry Maudslay.

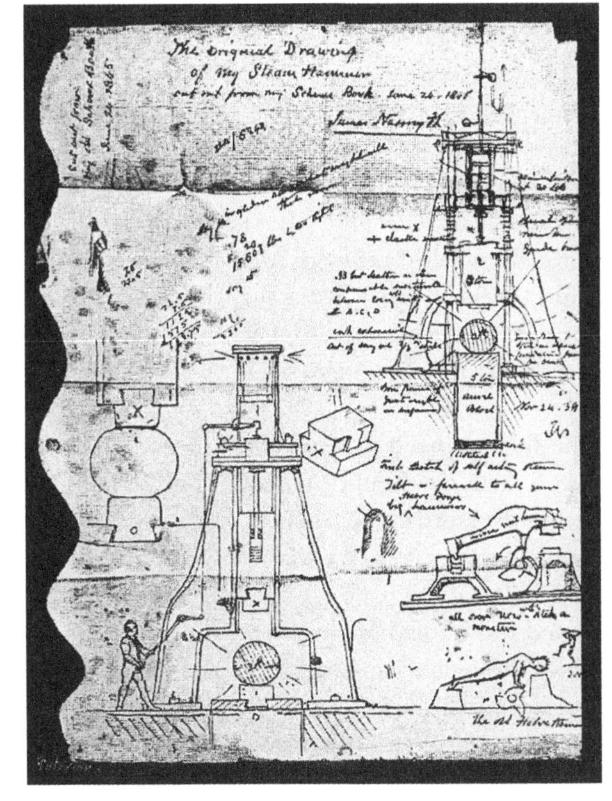

Nasmyth's original sketch for his steam hammer.

He looked back on those days with great affection, always referring to Maudslay as his 'dear old master'. By 1834, still only 26 years old, he felt confident enough to set up in business on his own in Manchester. He was so successful that in 1836 he was able to develop far bigger premises beside the Bridgewater Canal, where he manufactured locomotives and machine tools. He was just the man to help with Humphreys' dilemma.

Innovation rarely comes from a standing start, appearing almost instantaneously. Most new ideas are developments and improvements on existing inventions, like James Watt's vital work on the steam engine, which came from trying to solve a problem with the earlier Newcomen engine. This was not the case with Nasmyth, who provided a wholly new invention, completely formed in theory, almost as soon as he heard from Humphreys. Nasmyth realised that the answer was to have a hammerhead that could fall vertically onto the anvil. All that was needed was a force to raise it high enough and gravity would do the rest. He described what happened next in his autobiography:

> Following up this idea, I got out my 'Scheme Book', on the pages of which I generally *thought out*, with the aid of pen and pencil, such mechanical adaptations as I had conceived in my mind, and was thereby enabled to render them visible. I then rapidly sketched out my Steam Hammer, having it clearly before me in my mind's eye. In little more than half an hour after receiving Mr Humphries' [*sic*] letter narrating his unlooked-for difficulty, I had the whole contrivance in all its executant detail, before me in a page of my Scheme Book … Rude and rapidly sketched out as it was, this, my first delineation of the steam hammer, will be found to comprise all the essential elements of the invention.

The hammerhead was held within guide bars and was connected to a piston in an overhead cylinder. Steam was admitted to the cylinder below the piston, raising it up. Then an exhaust valve was opened and the hammer dropped under gravity. Later it was improved by allowing steam to act on both sides of the piston, providing extra impetus to the down stroke. It solved the immediate problem, and was soon adopted for engineering works throughout the world. The one job it was never called on to perform was forging the crankshaft for the *Great Britain*.

The paddle wheels of a steamer push water towards the stern, and the laws of physics decree that every action has an equal and opposite reaction. So, as the water goes back, the ship goes forward. But the paddle wheel is not the only device that can be used to move water. The Archimedes screw has, as its name suggests, been known since ancient times. It was mainly used as a helical screw inside a cylinder: as the screw was turned, water was drawn up the screw. Examples can still be seen, for instance, attached to windmills in the Netherlands for land drainage or, in a more domestic, modern setting, in the chocolate fountain. It is logical that if you set an Archimedean screw on a floating platform, stick the end under water and turn it, then the water will be pushed back, just as it is by paddles, and the platform will move. Two nineteenth-century engineers began experimenting with screw propulsion: John Ericsson of Sweden and a British farmer, Francis Pettit Smith. It was the latter whose work was to inspire Brunel.

Smith began his experiments on the homely setting of the farm duck pond. He tried a lot of different versions, each one powered by a simple clockwork motor. As the Archimedean screw works best with a large number of turns he must have thought that the same would apply to the screws for his model boats. Then, one day, there was a happy accident: the screw broke – and the broken screw worked better than any of the others. He had invented the screw propeller.

Encouraged by his initial success, he moved to a much larger experiment, fixing his new device to a steam launch that, with some justifiable pride, he named the *Francis Smith*. When that worked he moved up the scale and fitted out a three-masted schooner, which was equally appropriately named *Archimedes*. It was sent out on a test voyage round the coast, and inevitably it attracted the attention of the engineer ever on the lookout for the latest technological breakthrough, Isambard Brunel. He was convinced that the screw was a far better way of moving a large ship than paddle wheels, and he passed on the news to the unhappy Humphreys that his wonderful engine would not be needed after all. It was a dreadful blow and, according to Nasmyth, 'The labour and anxiety which he had already undergone, and perhaps the disappointment of his hopes, proved too much for him; and a brain fever carried him off after a few days' illness.'

The change from paddle wheels to screw propeller was a tragedy for Humphreys, but it marked a vital stage in the development of the iron steamship. The elements had all come together: large size, steam engine, iron hull and screw propeller. There were still yards turning out wooden sailing ships and paddle steamers, but more and more were following the lead set by the *Great Britain*.

There would be major changes, with improved methods of constructing the hull and far more efficient engines. Brunel's engines were immense, with four cylinders arranged in pairs as inverted Vs, each 88in in diameter. Size was essential because steam pressure was very low – a modest 5psi – and the shaft was turned through a cumbersome arrangement of chains. Later engines were worked with higher and higher pressures, and in order to make maximum use of the steam, instead of simply allowing it to blow away in the exhaust, engines were compounded – so that the steam would pass from the first, comparatively high-pressure cylinder straight into a larger low-pressure cylinder. Eventually, the most efficient system used three cylinders – a triple-expansion engine – and the steam would then be condensed as it exhausted from the third cylinder, providing pure water to be fed back into the boiler.

The changes revolutionised the entire industry. For untold generations, shipbuilding had been all about craftsmen working in wood. A complex system had developed in which the master shipwrights who designed the vessels worked to their own closely guarded plans, known as 'mysteries'. Apprentices who came into the industry were sworn to secrecy.

Part of a riveted hull on a vessel at the Scottish Maritime Museum, Irvine.

Alexander Stephen, who was later to go on to establish his own yard on the Clyde, was apprenticed in 1787 and the agreement with the master shipwright contained this clause:

> I do hereby oblige my self to pay the fee of three pound sterling money for teaching me the art of ship drafting as you produce it yourself, and one half when entered to said drafting and the other half when I can lay down a draft by my self. I also bind my self to teach no other person the same under the fine of ten pound sterling.

That fine would be the equivalent of several hundred pounds at today's prices – a definite incentive for a young man just starting out in his career to keep all the information he gained to himself.

This whole system of passing information down through the generations via the practical lessons of long apprenticeships meant that the men saw themselves as true craftsmen who deserved both respect and a decent wage for their hard-won skills. But as the shipyards began to turn from wooden sailing ships to iron steamers, new men appeared with very different skills – men who understood engineering and working with iron.

As the value of the new skills rose, so the value of the old diminished and the change was bitterly resented. The conflict between the shipwrights and the iron workers was to give rise to demarcation disputes and arguments over who should handle which part of the work. To the outsider they could seem absurd. To the men, the issue was as much about status as it was about pay: the shipwrights felt particularly aggrieved, as they had always been the highest paid workers in the yard. Now their pay was slipping behind that of others and there was, in their view, an absolute necessity to protect what they had against all comers.

The result was a system of almost Byzantine complexity. When, for example, in the 1890s, there was an argument about who should do the necessary woodwork in the storeroom of a ship being built on the Tyne, it was eventually agreed that it should be done by the joiners, unless there was any wood greater than 1½in, in which case it would be the shipwrights' job. One can only imagine the confusion this caused in practice, as work stopped while pieces of wood were measured and were then passed from one group of workmen to another.

There should have been no such problem when it came to ironwork, simply because there were no precedents to follow, no ancient rights to be upheld. There were, however, still difficulties in reconciling the old ideas about how a ship should be built and then adapting them to work with the new material. Old techniques were still used. At first, it was rather like the situation that Abraham Darby had faced when designing the famous iron bridge. No one was sure how to use the new material, so he had treated each iron segment as if it had been made of wood. The iron shipbuilders now did something very similar, scarfing joints together from pieces of angle iron to create the necessary shape.

The shipbuilder John Scott Russell, writing in 1865, described the situation in those days:

> It was truly curious to notice the early ship-builder … most assiduously putting together short pieces of angle iron, and regularly scarfing their ends together to form from their assemblage a single iron frame, all formed out of little pieces. Happily there came an end of this, for he found it easier, cheaper, and stronger, to have his frame rolled in one piece or at the most in two: and in a short time he found out how, out of an angle iron of moderate dimensions to bend or weld a complete frame, all in one length. But it was a long time before he could reconcile his mind to this fine simplicity; and even to this day there are rude ways of patching one angle iron onto the back of another, to

> give something like a rough imitation of the graduation in strength which used to run from keel to gunwale, in a timber ship.

The first stage in this new process was, as it had always been, to produce a design. Once the draughtsmen had produced the necessary drawings these were taken to the mould loft where they were converted into chalk outlines on the floor, which were then used to 'screive' – gouge the patterns into wooden boards. These boards were then taken to the floor below where the ship's frames would be bent.

The process was quite complex, but in essence, the screived boards were used to create a corresponding pattern of pegs set into a block. The iron was then swung by crane onto the block and bent and hammered to the correct shape. After that, holes had to be punched in the iron to take rivets. Even this is not quite as simple as it might seem.

One man who was among the first to try and study working with iron in a scientific way was William Fairbairn, who had advised Stephenson on the Britannia Bridge. In 1865 he wrote a book, *Iron Ship Building*, in which he explained why, for example, a punch was better than a drill for creating a clean hole for a rivet. He recommended using a punching machine that used a punched card – an idea borrowed from weaving and the Jacquard loom – to control the punch. It was, he wrote, very much better than the usual method of using a sled-hammer 'in the hands of one whose muscular developments are generally in excess of his reflective functions'.

Once the frames were ready, they were taken to the berth where the keel had been laid and gradually the skeleton of the vessel grew. The next stage was to prepare and add the iron plates. They had to be cleaned and cut to size, then punched for riveting. By the latter part of the nineteenth century, it had been recognised that tapering rivets were better than straight and that they should be countersunk. As a result plates were marked CKTS (countersunk this side) or CKO (countersunk over). The plates then had to be chamfered or rolled, depending on how they were put together. They could either be overlapped or the ends could be butted together and strapped, a third smaller plate riveted behind the joint.

The work of attaching the plates to the frame was arduous and dangerous. First the plates had to be set in place and screwed up to align the riveting holes. It was not so bad when the men worked on the sides of the ship, but at the bow and stern they had to work under a great curved structure while often standing on a couple of planks suspended down the side.

Even in the early twentieth century, safety had a low priority. As part of an oral history project carried out at Greenock on the Clyde, researchers recorded the stories of former shipyard workers. Daniel Murray started work as a plater's boy in 1928 and his experience was typical of the time:

> Oh safety, there was no safety precautions in them days. You just worked, you had on whatever you had, just a cap, that's all you had just a cap on your head, Say in them days if anything struck you on the head your head split open. Many a time we were hurted that way, you know.
>
> Shoes? You never got anythin', nothin' in the shipyard. Nothin'. All we wore was a pair o' overalls an' eh, just yer jacket. Maybe a pullover and yer jacket on the top, that was all ye had. You never got any gloves or nothin' in them days to cover your hands or that. A' we had tae dae was we made a pair o' leathers wursel fur yer hands, ye know. You go a bit o' leather out and you cut, you slit a hole in it and you put your hand through it an' that gave ye a grip. If you hadn't them when ye' were shearing' the plates yer hands used tae be all cuts wi' the shears, they were very sharp. They were very sharp. They cut yer hands. Many a time yer hand was a' jaggy and cuts wi' it.

Yet hard as his life was, he recognised that of the riveters was even worse:

> And in below the bottom o' the ship you got 2 riveters and a holder on and a heater boy. And they worked as a squad. And many a time ye watched the riveters riveting all day just in their trousers, no shirt or nothin' on. The sweat was pourin' out them knockin' in rivets, thousands a day. And they used tae have a big pail beside then wi' water in it and meal. Meal fur tae make a drink, ye know, for they were thirsty. A riveter was a hard job, no doubt about it. They were worth a' the money, the riveters.

The advance of the iron steamer was remarkably rapid. An article in *The Practical Magazine* in 1874 quoted the following statistics for shipbuilding tonnage in Britain:

	Timber	Iron	Total
1850	120,895	12,800	133,695
1860	147,269	64,899	212,168
1868	161,472	208,101	369,573

It is noticeable that, although there was a steady and steep increase in shipbuilding as a whole, the bulk of the growth was in iron ships, rising from just under 10 per cent of the whole to more than half in less than two decades.

The world of shipbuilding was changing rapidly, but not in every quarter. The Royal Navy was notably slow to embrace the new technology. The engineers of the early nineteenth century were not much impressed by the Admiralty, and no one was less impressed than Brunel. During the Crimean War he had proposed constructing a small screw steamer with opening bows – in effect, it would have been an early example of a landing craft. He put the suggestion to Sir John Burgoyne, who was an enthusiast for the notion and passed the idea on to the naval authorities – and heard nothing. Brunel was not surprised:

> You assume that something has been done or is doing in the matter that I spoke to you about last month – did you not know that it has been brought within the withering influence of the Admiralty and that (of course) therefore, the curtain has dropped upon it and nothing has resulted? It would exercise the intellects of our acutest philosophers to investigate and discover what is the powerful agent which acts upon all matters brought within the range of the mere atmosphere of that department. They have an extraordinary supply of cold water and capacious and heavy extinguishers, but I was prepared for and proof against such coarse offensive measures. But they have an unlimited supply of some *negative* principle which seems to absorb and eliminate everything that approaches them.

Certainly the Admiralty showed itself reluctant to abandon its sailing navy of wooden walls for a steam navy of ironclads.

They did have reasons for resisting change. The earliest steamers had all been driven by paddle wheels with much of the machinery above the waterline, making them hopelessly vulnerable. And paddle wheels set amidships took up space that should have been available for guns to fire broadsides. So nothing was done.

But change was forced on the reluctant Sea Lords by a change in technology. In Nelson's navy, battles had been fought between vessels lobbing round shot or canister shot at each other from cannon. This was not very efficient, and the alternative was being developed: the gun firing shells. In order to achieve greater accuracy, the gun barrel was rifled – a spiral groove inside the barrel engaged with a flange on the shell to set it spinning. The problem was that cast-iron guns tended to crack as the shell accelerated on its way. Steel was not yet available in any sort of quantity and was expensive, which only left wrought iron, which seemed even less likely to produce a satisfactory result. The answer was found by William Armstrong.

Armstrong was born in 1810 and, as the son of a successful businessman, was expected to go into one of the older professions, and law was the one selected. But he had always shown a keen interest in all things mechanical and the Bar was deserted in favour of forge and foundry. His first great breakthrough came with the development of hydraulic machinery, especially cranes for use in the country's docks and harbours. He set up a manufacturing base at Elswick, on the Tyne above Newcastle. It was the advent of the Crimean War that got him thinking about armaments. The existing heavy guns kept getting bogged down in the mud, and he began working out how he could get the same firepower from a lighter gun.

His system involved heating a wrought-iron strip, bending it into a spiral, then raising it to a white heat and hammering it to weld the spiral together to form a tube. A second, larger tube was then welded over the first. In order to rifle the interior he used a thin steel inner sleeve. The actual rifling proved difficult, but the job was carried out overnight at Elswick, with Armstrong staying up to see the process through.

His first 18-pounder was demonstrated in 1855, but it took another three years for the military to accept the new weapon. A demonstration firing against a conventional 18-pounder was certainly impressive. The Armstrong gun sent its projectile twice as far as the conventional gun and with far greater accuracy. It marked a new stage in warfare – and a new development for Armstrong, who turned his Elswick Works into the country's most important armaments factory.

The arrival of a more efficient gun made the old wooden battleships seem hopelessly vulnerable. The French were the first to react with *Gloire*, the first real attempt to create an armour-plated battleship, designed by Depuy de Lôme. It was launched in 1859, but it was really only a conventional wooden ship with extra iron plates hung on the outside. That same year, Britain launched its new battleship, HMS *Victoria*, a three-masted wooden vessel with three rows of cast-iron cannon poking through the gun ports. If it had not been for two stubby funnels indicating that the vessel could use steam as well as sail, there was nothing to distinguish it from the vessels that fought at Trafalgar. It was clear to everyone that, in a contest between the *Gloire* and the *Victoria*, there could only be one winner.

The British needed a warship to match the *Gloire*, and the task of designing and building it went to Isaac Watts, chief constructor to the navy, and John Scott Russell. The latter was one of the new generation of shipbuilders who believed in using scientific experiments to improve design rather than blindly following tradition. He asked fundamental questions: ones that had never been asked before, because no one thought they needed to know the answers.

One of these questions seems very simple, but in testing it he found some surprising results – what happens to the water as a ship pushes its way through it? He set up a trough of water and then released more water at one end and saw a single wave travel the length of the trough. It was, he said, 'a most beautiful and extraordinary phenomenon'. Then, in trials using a barge towed down a canal, he discovered that the vessel pushed just such a wave in front of it, but if the bows caught up with the wave, then the vessel reared up and slowed. It was like slamming on the brakes. He used his experiments to improve hull design. This was a new, modern approach to shipbuilding. He was just the man to help find a novel solution to the problem of how to create a battleship to meet the needs of the new age of gunnery.

The vessel designed by Watts and Scott Russell was HMS *Warrior* and, in the navy parlance of the day, she was a frigate simply because she had all the guns on one deck. But this was no ordinary frigate. At the heart of the ship was an armoured box, the citadel that contained the guns, ammunition and the vital supplies for running the ship. Damage to the vulnerable bows and stern would not prevent the vessel acting as a fighting unit. There was one problem to be overcome, however. How was the citadel to be made safe from attack? Simply using metal plates for the hull was definitely not the answer. Tests showed that the new shells not only went through the plates, but also sent metal fragments flying in all directions that would have devastated the crew inside. The answer was to create a sandwich. The outer hull was of 2in thick timber. Inside that were 4½in thick metal plates and the final layer consisted of a backing of 18in thick teak.

In many ways, the vessel was a completely new type of warship, with her steam engines and armour plating. But she still had three masts and a full set of sails and the guns were arranged to fire broadsides, just as they had for centuries. Even so, this was a formidable weapon that at once made all earlier warships obsolete. Until *Warrior* appeared, ships of the line with their rows of guns (vessels like *Victory*) were the pride of the navy. *Colossus*, built in 1848, just eleven years before the keel of *Warrior* was laid, had cost £100,000 and was just such a ship. She was now sold off for a miserable

£6,178 10*s*. Fortunately, *Warrior* has survived and the hull is just as it was and one can still see the construction of the citadel. The rest of the vessel has been rebuilt and fitted out as she was when launched and can be seen in the context of two older warships, *Mary Rose* and *Victory*, at Portsmouth.

One anachronistic feature of the first armour-plated warship was soon remedied. John Ericsson had seemed destined to be a runner-up in life, having his locomotive beaten at Rainhill and having lost the race to develop the screw propeller to Pettit Smith, but then he achieved a first. Now settled in America, he designed a vessel with a rotating gun turret. *Monitor* had an iron hull with a very low free-board which made it unsuitable for anything much more than harbour duty, and on top of that was a circular gun turret, the whole described by a contemporary as 'looking like a cheese box on a raft'.

In 1862, as part of the Republican Navy, it engaged the Confederate wooden wall, the *Merrimack*, in a long-distance duel that lasted for six hours with very little effect. *Monitor*'s guns were not powerful enough, and *Merrimack*'s shot bounced off the ironwork. But, if it did nothing else it showed that a rotating turret worked and provided protection for the gun crew.

In Britain, Captain Cowper Coles became an enthusiast for, what he called, the cupola turret and after testing it on an old floating battery it was considered a success. In 1862 the Admiralty commissioned HMS *Prince Albert*, with four turrets, each holding a 9in gun. It was obvious that fire-power was increasing and that new battleships would need better protection than even that offered by *Warrior*. The answer would need to be found by the iron industry.

John Brown was born in Sheffield in 1810. His father was a builder, but the boy had no intention of following that trade, mechanics was his passion. He left school at 14 and took an apprenticeship with a company manufacturing files and cutlery. At the age of 21, his apprenticeship was over. His father bought him a new suit, gave him a sovereign and told him he was now on his own. He made his way to good effect, rising through the company ranks and eventually setting up in business for himself as a factor.

In 1848 he invented and patented the conical spring buffer for use on the railway – a device of which he was so proud that, when he was rich enough and grand enough, it featured in his coat of arms. He was now able to establish his own manufacturing company, the Atlas Steel & Spring Works in Sheffield. His ambitions did not stop there. He began to develop an interest in working in steel, a story that will appear later on in Chapter 9.

He travelled widely, and while in France he got permission to see the *Gloire*. He recognised the importance of armour plating, but was not impressed with how the iron plates were made by hammering. He was sure a far better result could be had by rolling. He does not seem to have been aware that the idea had already been put into practice by Beale's in nearby Rotherham. They had rolled plate for a floating battery built by Palmer's of Jarrow for use in the Crimean War. But Beale's had lost interest in the process by the time Brown started his process.

A local paper gave an account of the work needed to produce a wrought-iron plate, 20ft long, 4ft wide and 5in thick:

> The foreman of the forge comes upon the scene, strips off his coat, and sees that everything is right for the momentous operation. The foreman of the rolling department is also at his post. Seventy men gather under his orders. A strong chain is passed round the rolls, which are about 20 yards in front of the furnace, to aid in drawing the great heated mass from the fire … the forge [then] opens, and the loose fire-bricks on which the front of the plate has rested are pushed away, that it may be grasped on its fore-edge by enormous toothed tongs, which it requires three men to lift. Now all is ready. The great door of the furnace, about six feet-wide, rises. The tongs grasp the plate. The rolls turn, the men pull, and out shoots in an instant the great mass of iron. And here comes into play a minor, but very necessary piece of ingenuity. The moment the whole length of the plate rests upon its carriage, a mechanical contrivance releases the chain, and it is drawn no further. As quick as thought the chain is detached from the rolls, and the men, seizing each side of the carriage as if they were salamanders, run it rapidly up to the rolls. It passes through, and at once its ten inches of thickness are reduced to eight. The rolls are reversed and back comes the plate, this time losing only half an inch of its thickness … In all it passes eight times through the rolls. Though this is done in five minutes, there is no hurry, but the utmost precision and care. At each side, when the hot plate emerges, its surface is swept with brooms dipped in water, and sometimes buckets of water are dashed over it, to remove the scales, and give it a fair surface. It is curious to see how important is the water upon the mass of glowing iron. Great drops of water run hissing over its surface, and it does not appear to lose anything of its heat. [It is then moved by crane] and two immense rollers of nine tons each pass over it, to ensure its being perfectly straight as it cools. The cooling process occupies about 12 hours, and another crane then lifts the plate to the planing tables, where the edges are cut, squared and afterwards grooved.

In 1867 Brown produced the thickest and heaviest armour plate ever rolled; 20 tons of metal were rolled to create a plate 40ft by 4ft by 4½in within about a quarter of an hour of leaving the furnace. Brown declared himself proud of his 'brave workmen'. The company prospered, and in 1857 they were employing 200 men. By the time the great plate was rolled this had risen to a workforce of 2,000 and the turnover had risen from £3,000 a year to £1 million.

In time, the demands for ever larger rolled plates meant that Brown's needed a far more powerful engine to work the rollers. In 1905 they acquired a monster built by the local firm Davy Brothers. This three-cylinder engine stood 28ft high, weighed 420 tons and produced 12,000hp. An identical engine, made for Cammel's is now in the Kelham Island Museum in Sheffield, and perhaps the most remarkable thing about it is to see how it can reverse direction in an instant – an essential, as the plate would have to be passed from a set of rollers turning in one direction to the next set turning the opposite way.

The *Queen Mary* under construction at the John Brown shipyard on the Clyde. (Scottish Record Office)

Brown's were later to move into shipbuilding and became one of the most famous names on the Clyde. Theirs was the yard that saw the launch of such great ships as the *Queen Mary.* Today the yard is derelict and only the great Titan crane stands over the empty docks as a reminder of former glories. But throughout the nineteenth century, Britain came to lead the world in the manufacture of iron steam ships, both for peace and war. The great yards were employing machine tools on the grandest scale.

When the Institute of Mechanical Engineering held their annual meeting in Ireland in the 1880s members went to see the major redevelopment at the great Harland & Wolff yard at Belfast. There were machines that could punch holes through 1½in plate, and planing tools that could work on plates 28ft long and 7ft across. Most impressive of all were the giant rollers.

The ships they were producing were no less impressive. Two liners for the White Star fleet, *Teutonic* and *Majestic*, had saloons with crystal domes in the roof and fitted out with the most elaborate decoration, 'bas-relief figures of tritons and nymphs in gold and ivory gable around'. But, however ornate the decoration, liners such as this were monuments to the skills of engineers and ironworkers. Their success also relied on the experience of engineers and manufacturers, who had turned away from the rule-of-thumb traditions of the old timber and sail shipwrights towards a more scientific approach, men such as William Fairbairn. But that was not the only work he did on the use of iron in construction: he also conducted experiments on using the metal in structures firmly based on the land.

8

THE IRON FRAME

For many people, wrought-iron balconies, decorative porches and railings epitomise Georgian Britain and especially the Regency period. Cheltenham, in particular, is noted for examples spread along its terraces and squares. Companies in the town specialised in this type of work. One of the more important, H.H. Martyn, advertised themselves as 'Sculptors, Carvers, Art Workers in Metal'.

But there was nothing new in the technology. The methods used had been practised by blacksmiths through the ages: heating, bending, hammering and twisting iron rods and welding sections together to make intricate patterns. The results could be extremely ornate, such as the very elaborate screen made by Cheltenham craftsman William Tetheren. This won him a gold medal at the Paris Exhibition of 1866, but a medieval craftsman, given a demand for such work, could have done the same. The truly revolutionary use of iron in buildings was not to be found in elegant terraces, but in mills and factories.

It was the development of new machines for the textile industry that blasted industrial change from a slow, gentle process to full-scale revolution. Work was moved from cottage to factory and the new buildings were on a scale that had previously only been seen in the great churches and cathedrals. They are instantly recognisable, multi-storey buildings with rows of windows to ensure there was good, even lighting for the work going on inside. The older buildings had their waterwheels, while the newer ones, built or adapted from the end of the eighteenth century, had their tall chimneys indicating the use of steam power.

The older mills also had something else in common: they were vulnerable to fire. In the very early days this was not necessarily due to industrial accidents. When Richard Arkwright tried to expand his cotton mills from Derbyshire into the traditional areas of Lancashire, he met fierce opposition from the existing body of textile workers, who saw that this would mean the end of the old ways of working, with wife and children carding and spinning and men weaving in their own homes.

Now the women and children were expected to move into the factories, where their lives were governed by the factory bell. As long as the waterwheel turned or the steam engine huffed and puffed then the machines had to be tended. To the manufacturers it represented progress, but to the workers it meant the loss of ancient rights and a new system they were sure they did not need. The *Annual Register* reported the events of October 1779:

> Manchester, Oct. 9. During the course of the week several mobs have assembled in different parts of the neighbourhood, and have done much mischief by destroying the engines for carding and spinning cotton wool ... In the neighbourhood of Chorley, the mob destroyed and burned the engines and buildings erected by Mr Arkwright at very great expence. Two thousand, or upwards, attacked a large building near the same place, on Sunday, from which they were repulsed, two rioters killed, and eight wounded; they returned strongly re-inforced on Monday, and destroyed a great number of buildings, with a vast quantity of machines for spinning cotton, &c. Sir George Saville arrived (with three companies of the York Militia) while the buildings were in flames.

Arkwright fully expected the mob to march on his mills in Derbyshire, and collected a formidable armoury of cannon, handguns and spears to repel them.

His partner in the cotton enterprise, Jedidiah Strutt, had a mill in Belper, Derbyshire, and he too prepared for the invasion of machine wreckers, turning the mill complex into a fortress with gun embrasures cut into the bridge that crossed the road between mill buildings. They are still there. The mob, however, never arrived. They were only concerned with keeping the new machinery out of Lancashire and had little interest in what was happening in Derbyshire.

Strutt's factories were unscathed, but one of his cotton spinning mills was burned down anyway, not by arsonists but in an accidental fire on the morning of 12 January 1803. It highlighted a problem that beset the early mills. The combination of hot machinery, oil and a highly flammable raw material meant that the fire risk was high. It was decided to rebuild the mill, but this time as a fireproof building. The idea was not entirely new; first steps had already been taken by William Strutt, son of the founder of the company, in the 1790s when he designed buildings using iron columns and floors of brick arches springing from wooden beams. The beams were then covered with plaster to make them fire resistant, if not fireproof.

Strutt was involved in the next major advance in mill architecture, and this part of the story begins in Leeds and then moves to Shrewsbury.

John Marshall was a manufacturer who helped develop the mechanisation of flax spinning, to create thread for the linen industry. He went into partnership with two brothers from Shrewsbury, Thomas and Benjamin Benyon, and built a mill in Leeds. The actual development of the machinery was largely the work of a local engineer who has appeared before in this story, Matthew Murray.

The Benyons were keen to start a similar business in Shrewsbury and the partners acquired a site, alongside the Shrewsbury Canal at Ditherington, on the outskirts of the town in 1796. Even before work on building the mill had begun, there was a devastating fire in their Leeds mill that must have set their minds firmly on ways of building the new mill to ensure it never suffered a similar fate.

The partners turned to a local man, Charles Bage, whose father had owned a paper mill and who had also entered into a partnership in an ironworks with, among others, Erasmus Darwin, a great dabbler in all things scientific and grandfather to the famous biologist. So, although Charles Bage earned most of his income as a wine merchant, with a little surveying work on the side, he also took a keen interest in the new developments in technology. He discussed iron bridge construction with Thomas Telford and had a long correspondence about fireproof buildings with William Strutt, who was a close personal friend. So when it came to devising the form of the new mill, he had a pretty good idea of exactly what was needed.

The result was the world's first completely iron-framed building. For the first time, both columns and beams were made from cast iron, with the floor supported, as it was in the earlier Strutt mill, on brick arches. Virtually no wood was used anywhere in the building; even the window frames and glazing bars were cast iron. It has stood the test of time and has recently undergone a complete refurbishment. It was not the perfect answer to the problem of creating a fireproof building as, although cast iron is strong in compression, it is less suitable for taking the stresses placed on a beam.

Now, when Strutt had to rebuild his North Mill at Belper following his fire he had a new model to work from, based on his collaboration with Bage. In some ways, it improved on the Ditherington original by incorporating wrought iron into the design as well as cast iron. The new building was to be the complete fireproof mill and it was widely admired.

Abraham Rees selected it to exemplify the very best in modern mill technology for his *Cyclopaedia* of 1812, and the German architect Frederick Schinkel, who visited in 1826, declared it 'the best in Britain'. Schinkel was to go on to use a similar technique for his own highly influential building, the Bauakademie in Berlin.

So, this is an important structure in the whole story of the structural use of iron and fortunately, unlike the Bauakademie, it is still with us. The building is L-shaped with the main range running parallel to the river, six storeys high and fifteen bays long, with a six-bay wing running at right angles to it. It is built of stone to the first floor level, then brick above that. It appears from the outside to be very similar to other conventional textile mills of the period.

Inside, however, is quite different. Here there are a series of very low-rise brick arches, which spring from iron beams supported on iron pillars. The beams themselves are, like the early fish-bellied rails, curved on their underside. The pillars are held together with wrought-iron ties to resist the thrust of the arches. Above the arches are brick floors. To see the basis of the structure you have to go down into the basement, which has something of the air of the crypt of a great cathedral. Here the cast-iron pillars are sturdy and sit on pyramids made up of massive stone blocks. Going up through the floors, the structure becomes more delicate, until it reaches the attic that was used as a schoolroom for the apprentice children who worked in the mill.

There are many variations on this basic pattern. The most attractive is unquestionably Stanley Mill, a woollen mill in King's Stanley near Stroud in Gloucestershire. Here, the pillar and arches have been turned into decorative colonnades.

One of the features of North Mill was the stove that blew hot air through the building via a system of vents, a primitive form of central heating. An interesting and ingenious variation was installed at Houldsworth's Cotton Mill in Glasgow, built just after North Mill in 1804–1805. Where the Belper mill was powered by a massive waterwheel, this one had a Boulton & Watt steam engine. The exhaust steam was passed through the hollow columns to provide the heating.

The Albert Dock, Liverpool.

The author came across an unusual modern adaptation of this idea in what was once America's leading cotton manufacturing town of Lowell, Massachusetts. Many of the old mills have been converted into other uses, one of which had residential accommodation on the upper levels and public spaces below. The hollow cast-iron supports here had been used as waste pipes, so that while sitting listening to a lecture one had the slightly alarming experience of hearing a sudden torrent of water rushing past you down a column as someone upstairs pulled a bathroom plug!

Once the system of building using iron frameworks was established it spread to other types of structure, notably warehouses. Among the best examples are those of the Albert Dock complex in Liverpool. During the eighteenth century, ships unloaded their cargos onto open quays, where they were checked by Customs & Excise and duty was paid on the spot. There were two major disadvantages to the system. First, pilfering was endemic, as it was all too easy to steal from the goods piled on an open dock, and second, ships could only be unloaded when there were customs men present. By the beginning of the next century the government had passed a law that allowed for goods to be stored in bonded warehouses and duty only had to be paid when the goods left the warehouse. Now docks could be closed in by the new warehouses and goods stored in safety as soon as they left the hold of the ship being unloaded – and there were no more delays waiting for an excise man.

One of the first closed docks was St Katharine's Dock in London, the work of engineer Thomas Telford and architect Philip Hardwick. In 1839, officials from Liverpool went down to London to see St Katharine's and borrowed many of the ideas they found there. As a result they determined to build a very large dock, with warehouses on all four sides, that was eventually to be named Albert Dock.

The engineer in Liverpool was Jesse Hartley. He was a civil engineer who, in 1824 at the age of 36, was appointed deputy surveyor for Liverpool docks, and he had hardly got his feet under the desk before the surveyor resigned and he was the man in charge. He was to remain in charge for the next thirty-six years, overseeing the great expansion of the port, and only made his final visit to his office three days before his death.

Although the principles of fireproof buildings had already been well established, Hartley wanted to make sure that for this huge project he had not only the best but also the most economic system available. He produced six alternative designs for fireproof and semi-fireproof warehouses and tested the different designs to see just how efficient they were. He made models to test their strength and also to test their fire resistance.

There were two basic versions: one similar to that already used in mills and another in which wooden beams and columns were sheathed in iron. He built a special structure 18ft square by 10ft high, and in that he placed his models together with a splendidly inflammatory mixture – eight tar barrels, three barrels of pitch and a ton or so of dry wood. The iron clad timbers survived for three-quarters of an hour, but once the heat got to the wood the whole lot went up in flames. Hartley went for the conventional structure of iron and brick.

The warehouses were built over the quays, the upper overhanging storeys supported by immense iron pillars. Iron was everywhere: in the cranes built into the warehouse walls, in the swing bridge that crossed the entrance to the closed dock and in the massive iron bollards that lined the quay. What is not apparent to the visitor is that the roofs of the warehouses are curved and covered with wrought-iron plates.

The dock office at the entrance was designed by Philip Hardwick and given an imposing Tuscan portico and pediment that, in spite of its classical appearance, is also made out of cast iron. Today, part of the dock complex has a very appropriate new function as home to the excellent Merseyside Maritime Museum.

All the buildings discussed so far have one thing in common: there is nothing on the outside to indicate that they are anything other than conventional structures. The outer walls of brick and stone give no visual clue to the actual iron framework. An early exception is the boathouse at the naval dockyard at Sheerness. It was the work of Colonel G.T. Greene, whose title perhaps gives a hint to his way of thinking. He was the director of engineering and architectural works at the Admiralty from 1850 to 1864. The building, completed in 1860, certainly is as much about one discipline as the other, but as in his title, engineering comes first. The iron frame is obvious from the outside in strong horizontals and verticals and, looking at it, it could easily be mistaken for a modern frame building of the 1960s. But it is very much not the accepted pattern for Victorian mills and warehouses.

One type of building was developed on Merseyside that used iron in a very different way. The Mersey Iron Foundry had a speciality – constructing what we would now call flat-pack buildings made entirely of iron. These were mostly churches, sent out to missionaries in the further reaches of the Empire (though whether anyone really enjoyed a service in a cast-iron building in equatorial Africa is doubtful). The company dominated an area that came to be known as 'the Iron Shore'. Here, they built a village for the workers, complete with a church, inevitably made of cast iron. The locals found the whole thing an embarrassment, churches were not supposed to look like that in respectable English communities, so the building was given a cosmetic skin of bricks. It is still there today.

There were, however, other structures being developed where the iron was not only obvious but was used to enhance the appearance of the whole structure. The eighteenth century saw a literally growing enthusiasm for exotic fruit and flowering plants and the development of greenhouses and conservatories. It was gradually realised that the growing could be greatly improved by increasing the amount of glass, especially in the roof. It was also recognised that, by using narrow cast-iron bars to hold the glass in place instead of much wider wooden glazing bars, the area of glass could be increased.

The earliest surviving example is probably the greenhouse at Chiselhampton in Oxfordshire, built around 1800. It backs on to a stone wall, but the other sides and the umbrella-shaped roof are completely glazed. The cast-iron uprights have a moulded decorative pattern of vines. The designer is unknown, but it was clearly someone with some basic knowledge of engineering and an ingenious mind. At the apex of the roof is a little iron and glass cap that can be raised to let hot air out and improve ventilation. It is counterbalanced by an iron ball, so the whole cap can easily be moved using a simple rope pull. Over the years, greenhouses and hot houses became ever more elaborate.

The most famous of the iron-framed glasshouses is undoubtedly the Palm House at Kew, where the effect is to create a vast open space, but one richly decorated with wrought-iron motifs and exotic spiral staircases. However, for pure simplicity, where the entire effect is to create an elegant space of beautiful curves, there are few structures to match the Kibble Palace in the Botanic Gardens in Glasgow.

John Kibble, an engineer-architect, had built a beautifully curved conservatory at Coulport on Loch Long, an inlet that bites deep into Scotland's west coast. The iron frame was specially manufactured by James Boyd of Paisley and was completed in 1865. Kibble discovered that his family did not share his enthusiasm, either for tropical plants or iron and glass structures, so he arranged to have the whole thing moved to Glasgow at his own expense. There, it was enlarged with an immense central area, 146ft across and covered by a double dome. The upper dome is carried on twelve barley-sugar twisted columns ending in filigree brackets and a further twenty-four columns support the lower dome. At the junction of the two domes is a clerestory with ventilators. The effect is of a structure that seems almost to float above the ground.

One of the masters of constructing giant greenhouses and conservatories was Joseph Paxton. His was a most remarkable career. A farmer's son, he was born either in 1801 or, more probably, 1803 and received little formal education. In 1823 he went to work at the Royal Horticultural Society's grounds at Kew, close to the Duke of Devonshire's London estate at Chiswick House. The duke was an enthusiastic gardener and often wandered round Kew to see what was new in the world of botany. He chatted to the gardeners and found young Paxton to be remarkably intelligent and informative, and he was so impressed that, in 1826, he brought him to Chatsworth, his main estate in Derbyshire, as head gardener.

To satisfy the duke's enthusiasm for raising exotic and tender plants, Paxton, who had absolutely no training in either engineering or architecture, designed a whole range of buildings including a lily house in which he managed to grow a giant lily from seed, a process that had defeated the professionals at Kew. His most ambitious building was the Great Conservatory that covered an acre of ground and had an avenue through the middle along which visitors could drive in a horse and carriage.

The relationship between the duke and Paxton was more like that of partners in an enterprise than master and servant, and Paxton received due recognition for his work – and made a considerable amount of money. His name might have remained known only to a few with a special interest in garden buildings, but for an almost accidental event.

In 1849 plans were being laid, with the enthusiastic support of Prince Albert, for a Great Exhibition to be held in Hyde Park. A competition was held to select an appropriate design for the exhibition hall. Altogether, 245 entries were received and every single one was rejected. In despair the committee turned to the most enterprising and original engineer of the day, Isambard Brunel. What the great man produced was a horror; an enormous brick building topped by a huge iron dome. Apart from being generally considered hideously ugly, it was also impractical – there was very little natural light and no one had worked out how the vast quantity of bricks could be produced, let alone laid in the time available.

Paxton was invited by Edward Shipley Ellis MP, who was chairman of the Midland Railway in which Paxton had an interest, to visit the House of Commons in June 1850 to see a demonstration of a novel ventilation system that had been designed for the new building. It proved a disaster, being both inefficient and unbearably noisy. The two men discussed the fiasco afterwards and Paxton casually mentioned that another fiasco seemed to be developing over the Great Exhibition, and he rather wished he had submitted a design himself. Ellis at once said that it was not too late, but he would have to hurry – there was just a fortnight left before the Brunel plans were likely to be approved.

A few days later, during a Midland Railway meeting, Paxton sketched out his idea on a piece of pink blotting paper. It was for an iron-framed glass building, built up in tiers to a central nave with a huge semi-circular roof. It was, of course, a sketch for the Crystal Palace.

Once the plans were approved, the contract for construction went to Chance Brothers to supply the glass, and Fox & Henderson of Smethwick to supply and erect the building itself. They had to provide 3,300 iron supporting columns, almost as many hollow columns for drainage and 2,224 main girders. Altogether they were to supply nearly 4,000 tons of cast iron and 700 tons of wrought iron. Everything had to be prefabricated off site then transported to London for erection. The *Official Popular Guide* published for the opening in 1851 gave details of the process:

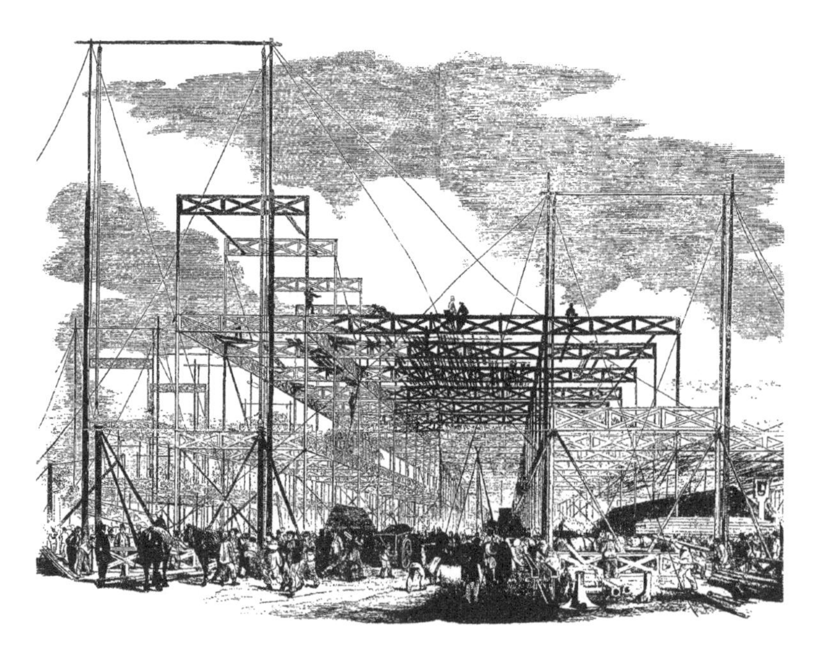

Building the Crystal Palace.

> As every casting was delivered on the ground, it underwent a careful examination, and was immediately painted. The girders, upon the perfect soundness of which the stability of the galleries and roof mainly depended, were subjected to a rigorous test, in a machine arranged for the purpose.
>
> A careful observation of this apparatus conveyed the assurance, that every girder, according to its ultimate destination, was proved to a strain of either 9, 15, or 22 tons. After testing, the girder was released from its confinements, again raised by the crane, and stacked in a convenient place ready for removal. So admirably were the various arrangements made for conducting these operations, that it was possible for a girder to be lifted from its waggon, weighed, secured in the testing-machine, proved, released, again raised, and finally deposited in less than four minutes.

Construction was kept very basic and simple. Each column had a connecting piece attached to the top. It was raised by shear legs and, when a second was in place, a girder was slotted in between the two. This was repeated until four columns were in place and joined by girders to create a rigid cell. This was then repeated over and over again to create the main framework. Once the frame was complete, the glaziers could get to work, using specially designed apparatus. In one week, eighty glaziers installed 18,000 panes of glass covering an area of over 60,000 square feet.

It was an immense undertaking. When work started in September 1850 there were just thirty-nine men employed. By the end of the year there were more than 2,000. But it was a success, and everything was completed on time ready for the royal opening on 1 May 1851. It is interesting to note that among the manufacturers who exhibited in the exhibition was the Coalbrookdale Company that had done such pioneering work in the iron industry. But where, in the early days, they had specialised in utilitarian items such as cooking pots and cylinders for steam engines, they were now showing off decorative ironwork, including the splendid ornamental fountain mentioned earlier.

In a sense, the Crystal Palace was a one-off, but the idea of a great glazed roof was to find another very much more common use: the new city railway stations. The earliest railway stations, such as that at the Manchester end of the pioneering Liverpool & Manchester line, were unassuming structures. The facade facing Liverpool Road has a vaguely classical appearance and could easily be mistaken for a company office block. No doubt the rather homely appearance was intended to reassure passengers that this new form of transport was really not that strange and was perfectly safe. Virtually the only indication that this was actually a passenger station was the platform at first floor level.

But as the system extended, so the demands on stations grew greater, and it was recognised that passengers needed protection from the elements and, in the case of the largest stations, it called for an overall train shed to cover all the platforms and tracks. The need was greatest for lines into London, and the first great terminus in the city utilising iron to the full was, almost inevitably, Paddington Station, Brunel's London terminus to the Great Western. Although the architect Digby Wyatt is credited with parts of the design, the overall concept is unquestionably that of the great engineer.

Brunel had lost out to Paxton in the competition to design the Great Exhibition hall, and had seen a cathedral-like structure of iron and glass being developed. Now he would create his own 'cathedral of steam' and, for once, the hyperbole is justified. The overall pattern actually mirrors that of a great church, complete with nave, side aisles and transepts. The great triple roof is supported by iron ribs, every third rib rests on an octagonal pier and the intermediate ribs on open cross-members. The central 'nave' has a 109ft span, with 70ft and 68ft aisles to either side. It remains one of the most thrilling buildings in London.

They are not the largest spans of the London stations, however, that honour belongs to St Pancras. The train shed there has been rather overshadowed by the exotic Victorian splendour of the recently refurbished Midland Grand Hotel – Gothic revival at its most picturesque. The station behind it, however, owes little to architectural fads and everything to the vision of the engineer W.H. Barlow. It has a single, soaring, slightly pointed arch with a span of 240ft. This great structure is held together by tie-bars hidden out of sight in the basement. This underground section was originally a goods shed, almost entirely devoted to the movement of beer barrels brought down from Burton-on-Trent.

There were other magnificent examples built throughout Britain, such as York and Newcastle stations, which have the same sense of drama.

By the end of the nineteenth century, more and more uses were being found for the attractive combination of iron frame and glass for light, airy spaces, combined with the possibilities of turning columns and girders into decorative features. It can be seen in, for example, the shopping arcades of Leeds and market halls such as the Grainger Market in Newcastle.

The ornate iron roof of the Grainger Market, Newcastle-upon-Tyne.

Iron that had once been thought of merely as a suitable material to reduce the fire hazard in industrial buildings where it could be safely tucked away out of sight was coming into its own. Now it could be used in even the grandest of structures.

The Royal Museum of Scotland was to be that country's response to the Crystal Palace and, once again, the design work fell not to a conventional architect but to naval architect Captain Francis Fowke, who began work in 1861. At its heart is a vast, glass canopied atrium, encircled by two storeys of iron balconies carried in iron piers. It is a triumph, but in many ways it comes at the end of an age. The next generation of iron buildings would be very different and based not on cast and wrought iron but on steel. However, they had to wait until the technology of steel production made production on a large scale possible.

9

THE BEGINNINGS OF A STEEL INDUSTRY

Today steel may be the commonest form of iron produced, but for centuries it represented only a small proportion of total output. Now it is used for a huge variety of manufactured products and in the construction industry, but centuries ago there was only one property that interested manufacturers – its ability to hold and keep a sharp cutting edge. And steel was being used in this way while Britain was still in the Iron Age.

Into the Middle Ages and beyond, the finest sword blades were made in the city of Damascus and later in Toledo. But the special steel from which they were made did not originate there. The discovery of how to make fine steel was made in India over 2,000 years ago, and is known as 'wootz' steel. The name first appeared in the English language in the eighteenth century, and is thought to be a corruption of the word 'ukku', meaning 'crucible' in the southern Indian language Kannada.

The steel was traded as ingots sent out from India to the Middle East and eventually found its way to Moorish Spain, hence the various names it collected en route. It was well known that a blade of wootz steel was both very sharp and flexible. Experiments have recently been carried out to see how a wootz blade works in action. Human heads are not generally available for slicing in half, but a pumpkin stands in well, and slow-motion film shows that the blade flexes to a remarkable degree on impact. It does the job and doesn't snap in the process.

What gives this steel its special characteristics? The ancients had no means of knowing exactly what it was that produced this valuable material, but modern research has been more helpful. The steel has bands of microscopic carbon and ferric carbide particles. The proportion of carbon is crucial. At 1.2 per cent the steel is quite brittle and would fracture on impact. At 1.5 per cent it is transformed, becoming superplastic: the crystalline structure can be stretched without shattering.

No one knows how many years of experiment went into discovering the secret in India around 300 BC, but we do have travellers' reports of the manufacturing process in later years. One of the most complete accounts can be found in *The Book of the Sword*, written in 1884 by the famous explorer Sir Richard Burton:

> About a pound weight of malleable iron, made from magnetic ore, is placed, minutely broken and moistened, in a crucible of refractory clay, together with finely chopped pieces of wood *Cassia auriculata* [a common Indian shrub]. It is packed without flux. The open pots are then covered with the green leaves of the *Ascelpias gigantean* [giant milkweed] or the *Convolvulus lanifolius,* and the tops are coated over with wet clay, which is sun-dried to hardness. Charcoal will not do as substitute for the green twigs.

Once charged, two dozen of the crucibles would be placed in a furnace, fired either by charcoal or dried dung and brought to a high temperature by bellows. After two or three hours, the crucibles were removed and broken open to reveal a lump of steel about the size of half an egg. The eggs were converted into bars by heating in a charcoal furnace 'not hot enough to melt them':

> The so-called Damascus blade was produced artificially, mostly by drawing out the steel into thin ribbons which were piled together and welded by the hammer. Oral tradition in India maintains that a small piece of either white or black haematite (or old Wootz) had to be included in each melt, and that a minimum of these elements must be present in the steel for the proper segregation of the micro carbides to take place.

We shall probably never know why these particular ingredients were considered essential, though it might be significant that both the cassia and milkweed were commonly used as herbal remedies. Perhaps all the early experiments were made using materials already known to have special properties. But whatever it was in the Indian system that produced such magnificent results, the secrets were not learned in the West. It is only one example of the fact that, in many respects, the countries of the Far East were far ahead of the West in terms of technology and science for very many centuries.

When steel manufacture began in Britain, the end product was considerably less sophisticated than the wootz steel made in the Far East. Producing steel in Europe was rather more of a hit and miss affair than it had been in India. There were three basic techniques used.

In the first, the smith would heat the ore directly in his hearth and hope that he got the right carbon content, which was certainly not guaranteed. The alternatives were to start with wrought iron and add carbon or do it the other way round – start with cast iron and remove carbon. One of the earliest accounts was written by Vanoccio Biringuccio, a metallurgist born in Siena in 1480. His classic work, *Pirotechnia*, was published in 1540, shortly after his death. Wrought iron from the bloomery was dipped into a bath of molten cast iron and kept there until it was decided that enough carbon had penetrated the bloom. It would then be treated by repeated hammering of the hot metal and quenching in cold water.

Another common process in use at this time, described by numerous authors, was the case hardening of iron objects – anything from suits of armour to files. This usually involved heating the objects in a closed container, mainly made of clay but sometimes iron, together with some material that would add carbon.

A fifteenth-century account by Giambattista della Porta from Naples described large quantities of files being packed into an iron chest with an unlikely mixture of soot, glass and salt, or sometimes with powdered horn. Samples were taken out from time to time to be tested for hardness. There were, it seems, as many different ways of producing steel or case-hardened iron as there were metallurgists, each of whom thought their methods the best and guarded their recipes in 'Books of Secrets'. Steel making at this stage was more of a mysterious craft than an industry.

The first reference in Britain to a true industrial method for making steel can be found in Robert Plot's *Natural History of Staffordshire* of 1686. In it he describes steel making in a cementation furnace. Bars of very pure wrought iron, mostly imported from Sweden, were packed together with charcoal into clay vessels that were then heated in the furnace for several days. The carbon from the charcoal diffuses into the outer layers of the bars, giving them a steel skin. When the bars emerged from the furnace they would be covered in blisters, hence the name 'blister' steel by which it was generally known. The blister steel could be improved by breaking it up into small pieces and repeating the heating process, so that the carbon was distributed more evenly throughout the metal, not just staying near the surface. It was now renamed 'shear' steel because it could be used for making shears for the woollen textile industry. For the finest quality, the whole process could be repeated yet again, the end product being 'double-shear' steel.

Because the best iron was imported from Sweden and arrived in north-east England, the cementation industry developed in that area. One early eighteenth-century cementation furnace has survived at Hamsterley in Durham. The main structure is a massive cone, built of rough stone blocks, with single storey extensions to either side. It is as imposing as any blast furnace and clearly shows that steel manufacture had entered into an industrial phase.

It is an interesting example of how apparently quite different industries are interconnected. When woollen cloth comes from the loom, it first needs to be cleaned and shrunk, a process known as 'fulling'. Then, in order to give the cloth a smooth finish, it has to go through two more processes. First the surface would be stroked with teazles to raise the nap and then it would be cut by heavy shears. This was a highly skilled job and the croppers who carried it out were among the highest paid workers in the industry. The nap had to be cut evenly and closely and an inadvertent cut could ruin a length. Success depended on the craftsmanship of the cropper – and the sharpness of the shears. A gentleman wearing an elegantly smooth coat could give thanks to both the cropper and the ironmaster.

One of the most important uses for steel has already been mentioned – the file. It is a humble object, yet many craftsmen would have been unable to carry out their work without a whole battery of these tools, from the coarsest to the finest. Whatever their size, they had to have one thing in common – they had to be hard enough so that the file removed metal from the object being shaped, not the other way round. Making files was an activity that remained unmechanised right through until at least the latter part of the nineteenth century. An article in the *Penny Magazine* supplement for March 1844 described the whole process in detail as it was carried out at the Marriott & Atkinson Works at Sheffield. It is worth quoting in some detail, just to give an indication of the complexity involved in producing such an apparently simple object.

A workplace within the factory consisted of the hearth and bellows, much like any other smithy, and a solid stone block, weighing about 3 tons, with one or more anvils on top:

> Except for the smallest files, there are two men employed at each forge – a striker and a forger, one of whom manages the fire, heats the steel, and acts as a general assistant; while the other is the superior workman, who hammers the file into shape, and is responsible for its quality. There are various notches, ridges, curvatures, and gauges, on and about his small steel anvil, which enable him to work the piece of steel into the proper form for a file, including the narrow handle, or 'tang'. The rate of working is such that at the whole of the sixteen 'hearths', about fifty thousand dozens of files are made in a year. Each man accustoms himself to the making of one particular size of file ... From the thickness and softness of the heated metal, there is very little rebound of the hammer, and this renders the work of the striker rather laborious, especially for large files, where a hammer of nearly twenty pounds is used.

At this stage the files were blanks, without the characteristic pattern of cuts. They were annealed to harden them and the edges ground. The blanks were now sent to a separate room, where file cutters were ranged down the sides in front of the windows to ensure a good light for the intricate work. Each man had his own bench and some form of vice to hold the blank. The cutters had special chisels, made of very hard steel, with edges designed to give a particular shape to each cut:

> The hammers employed (the heaviest of which weigh about nine pounds each) have the handles placed ... at such an angle that the cutter can, while making the blow pull the hammer in some degree towards him, and this gives a peculiarity to the shape of the tooth. If the tool is a flat one the cutter places the small steel tool on it at a particular angle, and with one hammer blow cuts an indentation. He then, by a minute and almost imperceptible movement, changes the place of the tool, and makes another cut parallel to, and a short distance from, the first; then a third, a fourth, and so on to the end of the file, shifting the file slightly in its fastening as he proceeds. Generally the file is cut doubly, one set of cuts crossing the other at an angle ... In this case he reverses the position in which he holds the cutting tool and proceeds as before. If the file be round or half-round, or have a curved surface of any kind, he still uses a straight-edged cutting tool; but as this can only make a short indentation, he has to go round the file by degrees, making several rows or ranges of cuts continuous one to another.

The task is not quite as straightforward as it sounds, since the angle of the cuts and the shape of the groove would vary depending on how the file was to be used. Once cut, the files would be hardened and in a final stage scrubbed clean by women, using sand and water. Each file was tested before despatch.

The author of the article was shown a 10in file, with both a curved and a flat surface, which had required a total of 22,000 cuts, each one made by hand and judged solely by eye. It is difficult to appreciate the craftsmanship in making a file unless one has actually tried it. I once had the opportunity at a forge in France and found it virtually impossible to achieve the essential regularity, making cuts that were either not parallel to the previous ones or not the regulation distance apart – or more often both.

Some of the finest files were used by clock and watchmakers. It was one of these who was responsible for the first real breakthrough in steel manufacture. Benjamin Huntsman was born in Lincolnshire in 1704, but after his apprenticeship he set up in business as a clockmaker in Doncaster. It is said that he became dissatisfied with the quality of the steel he could buy for use in springs and pendulums, and began experiments in 1742 to find a better system. He succeeded, and moved to Sheffield, not as a clockmaker, but as a steel manufacturer.

The process was basically very simple. He took blister steel and heated it in ceramic crucibles to such a high temperature that the metal became molten, something that had never been achieved before. This ensured that the carbon was evenly distributed throughout the steel, and it also meant that steel could be cast for the very first time. The process attracted a great deal of interest abroad, and brought industrial spies sniffing round the Huntsman Works.

The first to gain access was John Ludwig Robsahn, but he lacked one essential quality necessary in an industrial spy – he could not understand the process. The next visitor, Benkt Qvist Andersson, was a good deal more successful. He came to Britain in 1766–1767 and took the secrets home with him, beginning the first crucible steel manufacturing plant outside Britain at his works near Stockholm. Huntsman's process was hugely successful, and it was recorded that in 1873 Sheffield was producing 100,000 tons of steel a year from 3,000 crucible holes.

The process was later superseded, but one establishment has survived which shows the whole process from manufacturing the steel to turning it into scythe blades. This is the industrial hamlet of Abbeydale on the River Sheaf, now swallowed up by the spread of Sheffield, but originally enjoying a rural setting outside the town.

The work at Abbeydale began by making the crucibles themselves from fireproof clay. In the pot shop, the wet clay was spread on the floor where bare-footed men stomped up and down on it to ensure a consistent mix and to remove any air pockets or bubbles that might cause the crucible to burst in the heat of the furnace. The clay was then formed into the long thin crucibles.

Once they had dried out they were taken to the furnace room. The building is very distinctive, with a tall, rectangular chimney at one end to ensure a good draught to the fire. This was set under the floor and fed with coke to reach the high temperature of 1,600°C. Here the crucibles were charged, either with very pure iron and carefully measured quantities of charcoal or blister steel. Once charged, the crucibles were lowered through holes in the floor to the heart of the furnace, where they remained for four to five hours. Then they were removed and the molten metal was poured into moulds to create ingots.

Outside the furnace was a water trough, not to provide comfort for horses, but for the men who soaked sacks to wrap round their legs to provide at least some protection against the fierce heat. This was a dangerous place to work, with the intense heat and a floor constantly awash with water.

Lifting and pouring the crucibles was not easy. A special tool was used, rather like two large two-pronged forks set either side of a metal ring. It was a two-man job. Each man held a set of prongs and secured the ring round the crucible, which could then be lifted, carried to the mould and tilted to allow the metal to flow.

Some years ago I made a radio programme about Abbeydale and spoke to an elderly man who had been an apprentice there many years before. He found the whole scene in the furnace room quite terrifying, and on the very first time he had to help lift a crucible he was so nervous he let his end slip. Molten metal hit the wet floor, solidified and shot up like a fusillade of bullets. There were a few moments of pandemonium, but it was a while before anyone realised that the lad was missing. He was eventually found almost a mile away, sat shivering on the river bank. He still carries little black marks where he was shot by steel bullets from that first day at the furnace. He got over the shock and went back to work.

The whole site relied on water-powered machinery. The extensive millpond occupies a major part of the site, and its first application appears at the next stage. Before the ingots could be worked into shape they had to go to the reheating furnace, which was coupled to a blowing engine worked

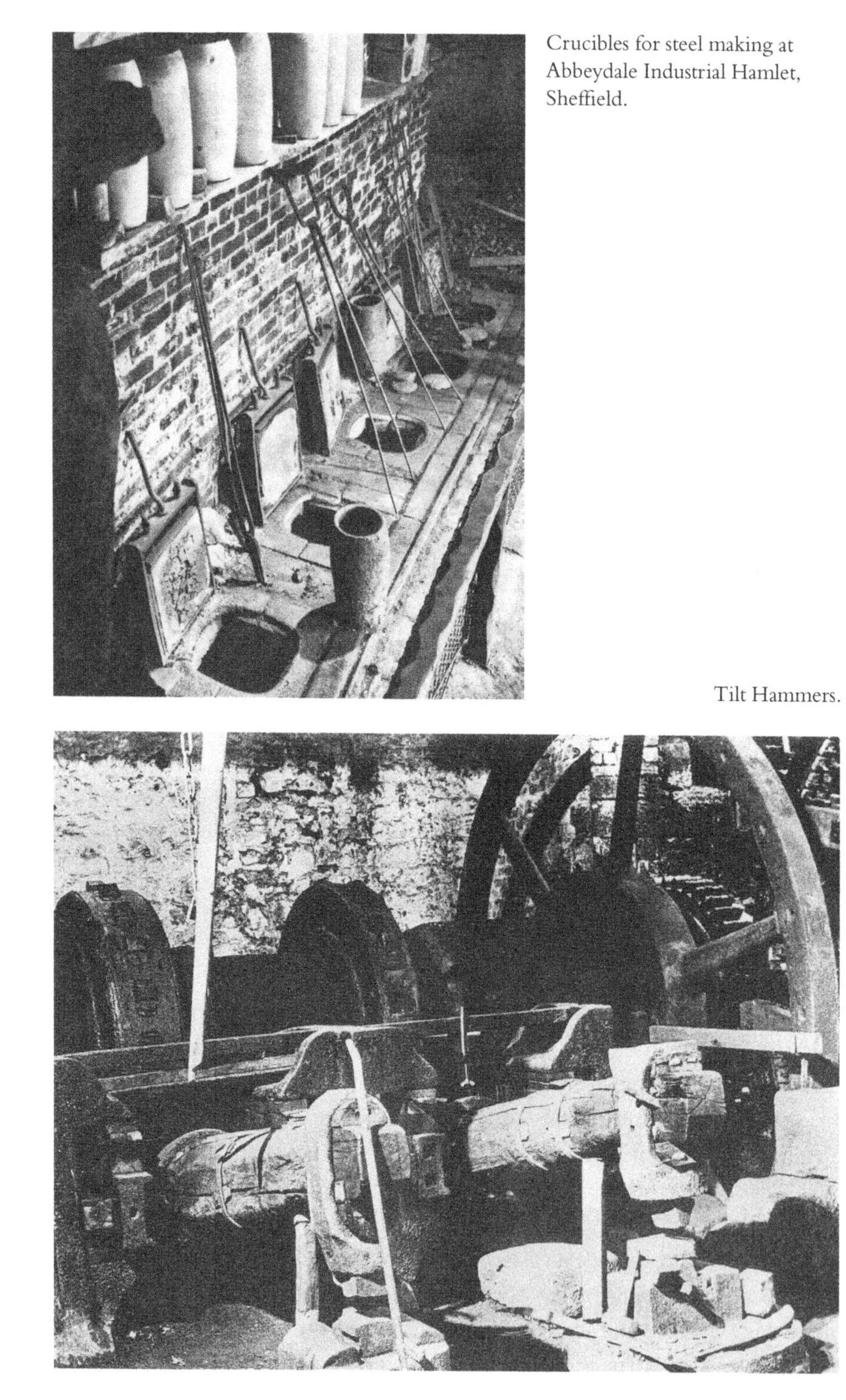

Crucibles for steel making at Abbeydale Industrial Hamlet, Sheffield.

Tilt Hammers.

by a waterwheel. This is one of those fascinating devices, which although basically quite simple are nevertheless very satisfying to see in action. An eccentric cam is rotated by a shaft from the wheel, and a smaller wheel runs over it, tracing out the irregular surface of the cam. This is attached to a piston and the movement of the little wheel makes it go up and down in a cylinder, blowing out air. (This is not only intriguing to watch – the movement is really quite hypnotic – it also emits a strange array of grunts, groans and whistles as it moves.)

The next stages were typical of forges of the period, in which the scythe blades were shaped using water-powered tilt hammers. In order to help them cope with the shock of the hammer blows, the workmen holding the metal sat in chairs suspended from the ceiling. And the force really was tremendous. The famous agriculturist Arthur Young visited a forge similar to this in his *Six Months Tour through the North of England* (2nd edition, 1770). According to him, the force was 'so prodigious: so great, that you cannot lay your hands upon a gate at three perches distance, without feeling a strong trembling motion'. As many of the families of the men who worked there

Grinding scythe blades in Sheffield in the nineteenth century.

also lived in the cottages on site, life must have been quite interesting when the hammers were at work, with crockery bouncing on the shelves.

Although Sheffield was the centre of the steel industry in the eighteenth and early nineteenth century, there were other parts of the country where agricultural tools were made for the local farming community. One of these sites is at Sticklepath in Devon, right on the northern edge of Dartmoor. It is called Finch's Foundry, an odd name, since there is no evidence it has ever been used for any form of casting. It is an interesting example of the adaptability of water power.

When William Finch came here in 1814 he took over a grain mill and an adjoining fulling mill to create his new works. The story starts on the River Taw above the town, where an artificial leat was cut to bring the water to the mills. It reaches Finch's on a launder, a wooden aqueduct, from which the water falls onto a succession of three overshot wheels. These are the most efficient form of waterwheel. Instead of water pushing at paddles, it falls into buckets on the rim, making one side of the wheel heavier than the other, forcing it to turn. Between them these wheels power all the machinery of the mill needed to produce agricultural tools, which were the main output, and tools for the china clay industry. When working flat out, five men could produce 400 hoes in each working day.

The last of the wheels to be reached by the water from the launder is the fan wheel, which does what its name suggests, works a fan that blows air through underground ducts into the hearths of the forge. There are three hearths providing hot metal for three sets of hammers, powered by the second waterwheel. Two are tilt hammers, working at different rates: one is governed by a tappet wheel with sixteen projections, giving sixteen blows per revolution, the other with twelve. The third hammer is a drop hammer, where the head is raised by a pulley system and then falls back onto the anvil.

The first impression you get when walking into the hammer room is that everything is a bit 'Heath Robinson', with frames held together by a motley array of struts and wedges. But in fact, everything is there for a purpose, and it is a reminder that places such as this were not designed on the drawing board, but put together by practical men using years of experience. The site is now in the care of the National Trust and, although visitors have to be kept at a safe distance, the machinery is still demonstrated.

When I first visited, the site was still privately owned and I was lucky enough to be able to join in the work of the forge and get at least an idea of what it was like to work with the giant hammers. It was a particular pleasure to have this opportunity, as my great-great-grandfather was foreman at a similar forge near Leeds in the nineteenth century. I am not sure what he would have made of my unskilled efforts to follow, if only briefly, in his footsteps.

The first task was to cut the hot metal into a suitable length, and it was surprising to find just how simple this was. A pair of large water-powered shears slices through the metal, and it really is as easy as the proverbial knife through butter. Then it comes to shaping on the anvil. Hammerheads were interchangeable, and a suitable head would be chosen for the job in hand. That bewildering array of wedges enables the main frame to be adjusted to change the length of the drop of the head. Starting and stopping the hammer is simple. When not in use, a solid hunk of timber is jammed between the hammer and the floor. Take it away and the wheel can turn and work gets under way; move it back in and everything stops again. As at Abbeydale, the hammering shakes the whole building and one is very conscious of the immense power generated by something as simple as falling water.

Originally, there was a woodworking shop for making handles, but that has gone. The final stage comes with grinding the edge of the tool. The grindstone is turned by the third wheel, and the operative lies on a contraption a bit like a diving board, holding the tool down onto the wheel spinning in front of him. When using this system you really do keep your nose to the grindstone. It feels uncomfortable and slightly dangerous to have your face so close to the wheel, where minute metal fragments are being worn away to create the cutting edge. It is not an illusion – this could be the most deadly part of the work of providing any form of cutting or sharp implement.

It is time to go back to Sheffield and Abbeydale to investigate further. Here, as in Devon, the edge tools finished up in the grinding hull, the name given to all grinding workshops in the region. Arthur Young had noted that the grinders were very highly paid: 18*s*–£1 a week in the 1770s:

> … but this height of wages is owing in a great measure to the danger of the employment; for the grinders; for the grindstones turn with such amazing velocity, that by the mere force of motion they now and then fly in pieces, and kill the men at work on them.

This was a very real danger. At the old forge mill at Redditch, where needles were made and sharpened, a fragment of a grindstone is embedded in the wall. There are initials carved on the stone; the initials of the man who was killed by the flying fragments. But although this was an obvious danger, it was not accidents that made grinding into a lethal trade; it was the stone dust and metal fragments that destroyed the lungs.

There was a broadsheet ballad on the subject, reprinted in *Poverty Knock*, one of the excellent collections of songs describing Britain's industrial past, collected and edited by Roy Palmer:

> The Sheffield grinder's a terrible blade,
> *Tally i o, the grinder!*
> He sets his little ones down to trade,
> *Tally i o, the grinder!*
> He turns his baby to grind in the hull
> Till his body is stunted, his eyes are dull
> And his brains are dizzy and dazed in his skull,
> *Tally i o, the grinder!*
>
> He shortens his life and he hastens his death,
> *Tally i o, the grinder!*
> Will drink steel dust in every breath;
> *Tally i o, the grinder!*
> Won't use a fan to run his wheel,
> Won't wash his hands ere he eats his meal,
> But dies as he lives, as hard as steel.
> *Tally i o, the grinder!*

The song does little more than hint at the misery suffered by many grinders. Dr G. Calvert Holland wrote a book, *Diseases of the Lungs from Mechanical Causes*, published in 1843, in which he identified what he called 'grinders' disease'. He did note, however, that workers in grinding hulls spread out along the River Sheaf, such as that at Abbeydale, fared better than many others for two reasons. First, they worked 'in the midst of scenery exquisitely picturesque and beautiful', which offered them the opportunity to enjoy a clean, healthy atmosphere outside the works. Second, they were mostly wet grinders. Their wheels turned in a trough of water that helped to reduce the dust, 'they are a fine healthy class of men, and have abundant means of securing the rational enjoyment of life … all the workmen and apprentices can both read and write'.

The modern visitor does still get a good impression of Abbeydale. It is notable that the terraced houses of the workmen are different only in scale from that of the owner and overseer. All are built in a sturdy, no-nonsense style of good, solid stone. In fact, the 'no-nonsense' theme even extends to the warehouse where the scythe blades were stored, ready for transport. The pillars are made out of old millstones piled one on top of each other. The builders followed the sound Yorkshire principle of 'waste not, want not'.

If places such as Abbeydale were considered healthy places to work, this was certainly not true of the great majority of the grinding hulls of Sheffield. Dr Holland was fierce in his condemnation:

> There is perhaps, no town in the united empire in which thoracic diseases prevail to so great an extent, among a large class of artisans, as in Sheffield. Grinding the various articles of cutlery and hardware, as we shall shortly explain, is an occupation peculiarly destructive to human life. The instances of suffering are not few and occasional, but numerous and constantly produced by the unmitigated evil of the occupation. Every practitioner here is more or less familiar with the disease induced; it is brought almost daily under his consideration.

Dr Holland backed up his bold statement with statistics. In the country as a whole, 16 per cent of working men died between the ages of 20 and 30; in the case of the fork grinders of Sheffield, that rose to 47.5 per cent. In other words, nearly half the grinders died before their 30th birthday, killed by the trade that earned them their living.

He was not content simply to publish the facts. Dr Holland campaigned hard to get something done to prevent the horrendous death rate, especially improving ventilation in the hulls. But twenty years later another doctor, J.C. Hall, found little had changed. In 1865, he wrote, 'A fork-grinder told me some years ago, "I shall be thirty-six next month, and you know that is getting an old man in our trade" ... I found the average age of the men only twenty-eight.'

Holland also looked at the needle grinders, concentrating on the area round Hathersage in Derbyshire. Here, he found a similar dismal story. 'The new hands are young men from 17–20 years of age, rough and uncultivated from the plough; and in those manufactories where ventilation is not secured, they are dead before the age of thirty, perhaps after two or three years of suffering.' They had a reputation among employers for staying off work at the least excuse and getting drunk. Who can blame them when they knew that they could be dead within a decade?

The workers in the town not only suffered worse conditions at work than their counterparts in places such as Abbeydale, they also endured far worse living conditions. Dr Holland looked at this aspect of life as well in another book, *The Vital Statistics of Sheffield* (1848). The typical cottages put up by speculative builders were three-storey buildings, consisting of a half-cellar day room, just 12ft square, with the main bedroom above that and an attic at the top. The speculators, according to Holland, 'never dream of the legitimate necessities of the population', but built increasingly flimsy houses, using tricks to give a false impression of solidity. For example, where a joist could be seen it might appear to be 2in thick, but where it was concealed under the plaster it had been cut away on the diagonal, so that only half the material was actually used, leaving the other half to be used in the same way elsewhere in the building.

The cutlery trade was not like other industries, such as textiles, where mill owners employed hundreds and even thousands of workers. Sheffield was a town of 'little mesters', each employing just a few hands. Attempts to improve conditions were hampered by many factors, including the multitude of small businesses, the laws governing trade unions that made it difficult to maintain any sort of solidarity and the proliferation of small unions – knife grinders would be in a different union from fork grinders, for example.

Unscrupulous union leaders, who were among the minority but nonetheless significant in number, abused the system. They resorted to all kinds of devices to force a man to join their particular union. The commonest method was 'rattening', which usually meant stealing the belt that drove his grindstone, making work impossible. If he paid his dues, the belt reappeared as mysteriously as it had vanished. But there were other, more violent methods used, including putting gunpowder into the chimney of a non-union man.

In 1864, William Christopher Leng came to Sheffield to run the *Sheffield Telegraph* and began a campaign to end what had become known as the 'Sheffield outrages'. As a result, a Royal Commission was set up in 1867 and offered protection and amnesties to anyone giving evidence. Soon witnesses began to come forward. A typical question and answer came during the evidence of T. Fearnehough:

> 'Has anything been said to you or done to you by the Saw Grinders' Union for working in this way, not being a Union man and not paying the sum of money?'
>
> 'Yes I got blown up for it.'

But by far the most dramatic evidence was provided by two men, Hallam and Crooke, regarding the murder of a saw grinder, James Linley. It appears that he had earned special hatred, not just for being non-union but also for taking on too many apprentices, reducing costs by means of paying the lowest wages.

There had been various attempts to intimidate him when the head of the Saw Grinders' Union, William Broadhead, approached Hallam offering him money to shoot Linley. There was a certain amount of haggling over the price, Hallam initially asking for £20 but eventually settling for £15. He recruited Crooke to join him in the enterprise. Those involved claimed they only meant to injure Linley, but he was shot in the head at close range while drinking in the Crown Inn. He died shortly afterwards.

Crooke admitted to being the one who actually fired the gun, simply because he was a better shot. Broadhead admitted his part, but said the rest of the union knew nothing about it and he had falsified the books to cover up his payment for intimidation. Because all the evidence was given under the promise of immunity from prosecution, no one was ever brought to court for Linley's murder. Broadhead, however, was disgraced and emigrated to America. Crooke went through a religious conversion and was eventually taken back to work by his old employers.

It was a dark interlude in Sheffield's history, and would probably never have occurred without the secrecy forced on the unions by the legislature. The law may have been designed to protect employers, but it provided a perfect cloak for the unscrupulous. The majority of the unions were never involved in the outrages, but there was undoubtedly widespread support for any movement that tried to change a lethal trade into one where a man could earn a decent wage without the shadow of death constantly falling over his shoulders.

10

THE TIN CAN

It might seem odd in a book about iron and steel to devote a chapter to a different metal, but it was never really a tin can at all – a more accurate name would be 'tinned' can.

The story begins with a problem that had been concerning mankind for centuries: how to preserve food that is produced seasonally for use in the rest of the year. Traditionally, there had been two common methods for the preservation of foods – salting and pickling – which served households well. There were others, however, who had problems, including soldiers campaigning abroad, where food was often scarce, and sailors on long voyages. The latter were particularly liable to suffer from a lack of fresh fruit and vegetables, ending with the seriously debilitating disease scurvy.

It was a problem that was faced by Napoleon's armed forces, and in 1795 his government offered a reward of 12,000 francs for a new, reliable method of preserving food for long periods of time. The franc had just been established as the new unit of currency and its value was set at 4.5g of fine silver so that, in theory at least, the prize was equivalent of receiving 54kg of silver, which was certainly worth taking a bit of trouble over.

One man assiduously applied himself to solving the problem of preservation. Nicholas Appert was born in Châlons-sur-Marne in 1749. His father was an innkeeper, but Appert moved to Paris to set up in business as a confectioner, though he also took an interest in other activities including that of a vintner. It may have been the latter that was his inspiration, as it was, of course, well known that wine in a stoppered bottle keeps its flavour and character but soon loses its quality when exposed to air.

Treating food was not quite that simple – simply putting food in a bottle and stoppering it did not work – and he spent fifteen years in experimenting before he solved the problem. Basically, he partially cooked the food and then bottled and loosely corked it. He then immersed the bottle in boiling water to expel the air and tightly sealed it to complete the process. There was no scientific theory on which to base his experiments, as Pasteur's work on bacteria still lay in the future.

Various commodities were bottled – this being France, the items included not just common vegetables but also delicacies such as partridge – and they were taken to sea by the French Navy for four months. At the end of the period, the bottles were opened and the contents were declared to be in perfect condition. Appert received his prize money in 1809, and the following year he published details of his method of food preservation in a book that was translated into a variety of different languages, including English. And it was in England that the next important developments occurred.

Peter Durand was a merchant who, although he admitted to having worked on the basis of information provided by a friend from abroad, received a patent in 1810 for preserving food. The method was basically the same as Appert's, with one crucial difference. Where Appert had only used bottling, Durand proposed using different types of containers, including ceramic and metal. There was an obvious disadvantage in having bottled food, especially if you were on a campaign with the army or navy – glass breaks.

Durand concentrated on metal. He sold his patent to Bryan Donkin, an engineer and inventor, who was later to achieve a certain amount of fame as developer of a rotary printing press. In 1813, Donkin went into partnership with John Hall to establish the very first canning factory at a site in Bermondsey. As with Appert's bottled food, the cans were tried by the navy and found to be satisfactory. Soon, any naval vessel setting out on a long voyage would be carrying a large quantity of canned food.

Among the most demanding of all the naval expeditions of the period were those undertaken by Sir William Edward Parry, who went to the Arctic in the search for the elusive north-west passage to the Pacific. One 4lb tin of roast veal survived from his journeys in the 1820s and was kept in a museum until 1938, when it was opened. Scientific tests showed that the meat was in reasonable condition and had lost little of its nutritive value. The experiment over, it was fed to a cat that was said to have scoffed it quite happily with no ill effects. It is a remarkable testimonial to the value of canned food.

Over the years, the methodology was improved. It was found, for example, that preservation was better at a higher temperature than that of boiling water, and in 1841 a patent was taken out for immersing the can in a bath of boiling calcium chloride solution. The process was an undoubted success, but only because the right material was available for making cans in the first place – tinplate.

Tinplate was first made in what was then Saxony, especially in the area round Dresden, in the seventeenth century. In its earliest crude form it consisted of iron, hammered out into a thin sheet and then coated in a thin layer of tin. It began to replace pewter for all kinds of domestic objects. The two materials shared common useful properties. They did not taint food, they did not rust and, unlike crockery, they were more or less unbreakable. Where tinplate had the advantage was in the all-important matter of cost, as it was very much cheaper.

It took a while for manufacture to arrive in Britain. The first tinplate works seem to have been set up at Pontypool in the 1720s. The industry relied on three main ingredients: iron, tin and fuel. Iron was readily available from furnaces in Wales, and the country had some of the most productive coalfields in Britain. Tin was the odd one out, as it came from mines in

Rolling sheet metal at a tinplate works.

A nineteenth-century canning factory.

Cornwall, but because the smelting process used more fuel than ore it made sense to bring the ore to the fuel rather than vice versa. Even before deep mining began in Wales, what was then known as 'sea coal' – coal that had been washed out of coastal deposits – was available. It was used for smelting Cornish copper as far back as the sixteenth century. South Wales was the obvious place to develop an industry where everything was available locally.

The invention of the tin can brought a new surge in demand and soon South Wales was established as the world's most important centre for tinplate production. By now, the original method of producing thin sheets by hammering had given way to the far more efficient system of hot rolling. By the middle of the nineteenth century, the work of rolling thin sheets of iron, and later of steel, relied entirely on the tinplate industry. A century after tinplate manufacture was first introduced into South Wales the area had eighteen mills at work rolling iron, using methods that remained more or less unchanged for another 100 years.

Essentially, it consisted of heating and reheating the metal in furnaces and then passing it through a succession of rollers that could be screwed closer together as the process required. The last of the old mills has long since stopped working, but fortunately not before the process was captured on film. In his book *Industrial Archaeology of Wales*, D. Morgan Rees quotes the commentary from that film describing the processes at the Teilo Works at Pontarddulais. It is one of the most complete accounts of the process that we have and forms the basis for the following description.

The power for the works was provided by a steam engine. By this time, steel had replaced iron as the basic material. It was brought to the site, cut to size and heated in the main furnace to a temperature of almost 800°C. The bars that came from the furnace, known as 'thick iron', were generally about 8in wide and ½in thick and could weigh anything up to 80lb each, so handling them by hand was hard work. After leaving the furnace, they were then passed on to individual work stations, each one consisting of two pairs of two-high rollers – roughing rollers and finishing rollers – and a smaller reheating furnace.

The two men at each station worked two bars at a time. Using special tongs, the first man passed a bar through the lower of the roughing rollers to the receiver on the other side. The receiver now passed the bar back through the upper rollers. At the same time, the first man started the second bar on its journey through the lower pair. After four passes, the bars were now known as short singles and were returned to the reheating furnace. Now the bars were passed through the finishing mill and after two passes, the short singles had become long singles. The long singles were now doubled over, reheated and again rolled, perhaps for two or three passes.

The process of doubling, reheating and rerolling continued, with a different name given at every stage. After the first pass of the doubled bars, they became short doubles, then they became long doubles, which in turn become short fours, then long fours, short eights and finally long eights. At every stage, the iron was squeezed, the folds were cut with shears and the metal sheets were separated from time to time to prevent them sticking together.

Once the rolling process was completed, the hot sheets were annealed in wrought-iron pots, and then pickled to remove any scale produced by oxidation during the earlier processes. In the early days, a variety of different pickling agents were used, including acetic acid produced from stale beer. In later years, sulphuric acid became the first choice. The sheets were now ready to be moved on to the next process, coating in tin.

The process of tinning was described in an article by E.J. Trubshaw, published in the *Journal of the Iron and Steel Institute* in 1883. The plates were first immersed in palm oil, then moved to the tin pot, where they were immersed in a bath of the molten metal for a few minutes. They were then re-dipped, returned to the grease pot and passed through rollers to spread the tin evenly across the surface of the plates. They were finally cleaned with bran applied by sheepskin.

By the end of the nineteenth century, the process had been simplified and mechanised. While iron was still used, a distinction was made between iron that came from charcoal furnaces, that was considered superior, and iron from a coke-fired furnace. Converted into wrought iron by puddling, they were known as 'charcoal plates' and 'coke plates' respectively. Even when steel was introduced to replace iron, the same names were retained to describe the different quality of plates. The plates themselves were generally about 0.012in thick and the thinnest coatings of tin were as little as 0.00006in per side.

Britain developed a near monopoly of the industry in the nineteenth century. Output was measured in boxes, which usually worked out at twenty boxes to the ton, and in 1890 South Wales was turning out 14 million boxes a year, over half of which went to America. Then, in 1890, the McKinley Tariff slapped a tax of almost 10*s* a box on tinplate imported into America, and soon an industry developed on that side of the Atlantic, using the same technology as South Wales. Within just a few years the American industry was self-sufficient.

The tinplate was formed into cans which, in the early part of the nineteenth century, were handmade by tinsmiths. Sheets of tinplate were cut to size and then bent into a cylinder over a roller. The edges were overlapped and soldered together. Discs were cut to form the two ends, and one disc soldered onto the cylinder to create an open-topped can. The partially cooked food was placed in the can and the top soldered into place, with a tiny hole left in the lid. The can was then heated in the salt bath until virtually all the air had been expelled, after which the hole was soldered shut. It had one final test to undergo. The filled can was then heated. If any of the food had gone bad, it would have produced gases that would expand and blow the tin apart.

In later years, the testing was improved by heating to a higher degree under pressure. This might not be thought of as a dangerous operation, but a bizarre account appeared in the *Illustrated London News* in January 1852, as part of an article about a London cannery:

> We well remember having recited to us the relation of an operator who was killed most ridiculously and ignobly by a boiled turkey. The canister in which

> the bird had been soldered was exposed to this process of heating under pressure and steam was generated beyond the power of the canister to endure. As a natural consequence, the canister burst, the dead turkey sprang from his coffin of tinplate and killing the cook forthwith made him a candidate for a leaden one.

Canned food and, later, canned drinks became commonplace and the industry thrived. In 1883 Norton Brothers of Chicago introduced a semi-automatic machine, using mechanical soldering that could turn out 2,500 cans an hour, and this later improved to 6,000. The cheap can brought in the age of convenience foods, epitomised by Campbell's soups, first seen in 1899 and selling for just 10 cents.

The loss of the American market was a setback for South Wales, but the deathblow came with the introduction of the aluminium can in the twentieth century. The industry came to an end, but it is still possible to see the remains at Kidwelly, where a tinplate works, begun in 1801, is now open as a museum, easily recognised by the tall stacks above the tinning bays.

Tinplate has been relegated to industrial history. It was not to be the only casualty of advances in technology. Puddled wrought iron would also become a thing of the past, replaced by mild steel. As Andrew Carnegie, the great steel magnate of America, put it – the Age of Iron was dead: the Age of Steel had arrived.

11

THE NEW STEEL AGE

Although steel was available in modest quantities for edge tools and cutlery, it was still considered quite an exotic material throughout the eighteenth century, to such an extent that it was actually considered very fashionable to sport steel jewellery. Indeed, the great manufacturer and entrepreneur Matthew Boulton, famous for his association with James Watt as a manufacturer of steam engines, made it into one of his most profitable lines in his early career.

The properties of steel were well known and valued, so there was clearly a demand for the metal, if only someone could find a way of manufacturing it in large quantities and at a reasonable cost. In the nineteenth century, two new production methods were invented that completely revolutionised the industry and led to profound changes in the world at large.

Henry Bessemer was born at Charlton in Hertfordshire in 1813. His father, Anthony Bessemer, had pursued an interesting career. He had been born in London, but his parents had moved to the Netherlands where he trained as an engineer. He went to work in France, where he became a member of the Académie Royale des Sciences and worked as die-sinker and engraver at the Paris Mint. The family would no doubt have stayed in France but for the Revolution. As it was, they hurried back to England, where he established a type-founding business.

Like many of the practical men who moved technology forward, young Henry received only an elementary academic education, but an excellent practical one in his father's workshop. He was fascinated by machinery, spent hours studying the workings of the local grain mill and soon began his own experiments, developing his own slide-lathe while still a teenager.

When he was just 17, the family moved to London. The boy had no professional qualifications, but he did have an unbounded confidence in his own abilities.

In his autobiography, he wrote that nature had supplied him with 'an inventive turn of mind, and perhaps more than the usual amount of perseverance'. He was soon applying these talents to a range of inventions, some more successful than others. He soon learned, however, that inventing something and getting a due reward for the invention were very different matters. He devised a die stamp for use by the official government Stamp Office, which produced a pattern of perforations on documents making them very difficult to counterfeit and making it impossible to remove the official stamp and reuse it. The government accepted the idea with enthusiasm, but neither offered him a job nor paid him. He decided that, in future, he would have to make the most of his ideas himself.

In the 1840s he found a new method of producing bronze powder, which was used instead of actual gold to produce the effect of gilding on decorative objects. He set up his own workshops in his house, and made enough money to turn the house into a permanent factory and workshop and buy a new home for himself in Highgate. In 1834, he married and started a family and continued on his inventive way. Over the years he took out over 100 patents.

However, it was not to be ornamental bric-a-brac that were the inspiration for the project that would make his fame and fortune, but the instruments of war. Among the ideas that he had developed was one where artillery accuracy could be greatly improved by the use of a rifled barrel that would give spin to the shot as it left the gun. The British War Office, a notoriously conservative body, was totally uninterested and he took his invention to France instead. However, he realised that he faced a real difficulty. He could not find a way of producing a rifled barrel using either wrought or cast iron, and steel was not available in large enough quantities to make it a practical alternative. So he went back to the basic question: could he produce steel in large quantities using some new technique?

Bessemer gave a full account of his experiments at a meeting of the British Association for the Advancement of Science held at Cheltenham in August 1854, and it was considered such an important event that the entire paper was reproduced in the *Times* on 14 August. He began by describing how he had spent many frustrated months building and rebuilding furnaces and working with small amounts of iron before he began to sense he was on the right track. It was only after many small-scale trials that he felt confident enough to set out his first large-scale experiment with 7cwt of pig iron:

> The numerous observations I had made during this unpromising period all tended to confirm an entirely new view of the subject, which at that time forced itself upon my attention – viz, that I could produce a much more intense heat without any furnace or fuel than could be obtained by either of the modifications I had used, and consequently that I should not only avoid the injurious action of mineral fuel, but that I should at the same time avoid also the expense of the fuel.

The whole basis of his work was the fact that he had realised that 'carbon cannot exist at a white heat in the presence of oxygen without uniting therewith and producing combustion'.

Bessemer built a cupola furnace, 3ft diameter and 5ft high, with five tuyeres near the bottom through which air would be blown. A hole at the top allowed molten iron to be poured in, and a hole near the bottom, closed by a plug, would allow the metal to be tapped. The process was started by blowing in the strong blast of air to prevent the metal escaping through the tuyeres.

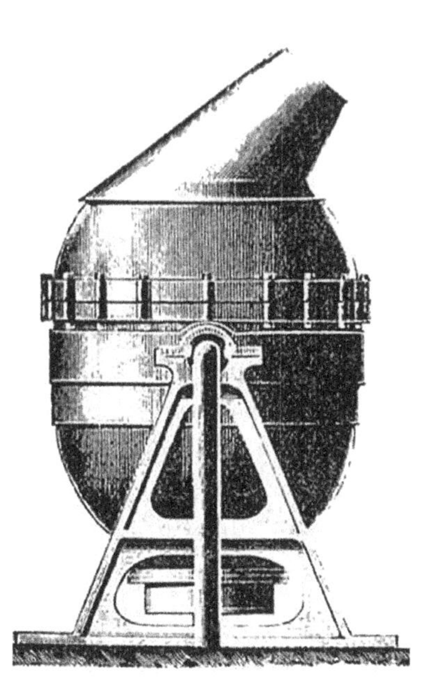

The Bessemer Converter.

Then the metal was added, and Bessemer described in detail exactly what happened. The language is undemonstrative, but it must have been an extraordinary experience to witness this violent process with its accompanying pyrotechnics. The account starts with the addition of the iron to the furnace:

> This having been done, and the fluid iron run in, a rapid boiling up of the metal will be heard going on within the vessel, the metal being tossed violently about and dashed from side to side, shaking the vessel by the force with which it moves, from the throat of the converting vessel. Flames will then immediately appear, accompanied by a few bright sparks. This state of things will continue for about 15 or 20 minutes, during which time the oxygen in the atmospheric air combines with the carbon contained in the iron, producing carbonic acid gas, and at the same time evolving a powerful heat. Now, as this heat is generated in the interior of, and is diffusive in innumerable fiery bubbles through the whole fluid mass, the metal absorbs the greater part of it, and its temperature becomes immensely increased, and by the expiration of the 15 or 20 minutes before named that part of the carbon which appears mechanically mixed and diffused through the crude iron has been entirely consumed. The temperature, however, is so high that the chemically combined carbon now begins to separate from the metal, as is at once indicated by an immense increase in the volume of flame issuing out of the throat of the vessel. The metal in the vessel now rises several inches above its natural level, and a light frothy slag makes its appearance and is thrown out in large foam-like masses. This violent eruption of cinder generally lasts about five or six minutes, when all further appearance of it ceases, a steady and powerful flame replacing the shower of sparks and cinder which always accompanies the boil.

Bessemer went on to explain that, due to the high temperature, impurities such as sulphur were oxidised and driven off. Once the flame at the top of the furnace died away, the process was complete and the hot metal could be run off.

What Bessemer did not tell the audience at Cheltenham, and what we would dearly like to know, is what his reactions were when he tried the experiment for the first time. It must have been alarming: the boiling of the iron and the juddering of the furnace, the flames and sparks and the final volcanic eruption of cinder, before the flames finally died away.

Bessemer took out a patent and invited industrialists to pay for setting up his system. Within days of the Cheltenham paper being published several large companies took out licences, including two of the most important concerns in South Wales, Dowlais and Ebbw Vale Ironworks. Bessemer was paid £27,000 and looked forward to a far greater fortune once his process became more generally accepted. He was to be hugely disappointed. Far from confirming his own success, the companies found that, even though they had followed exactly the same procedures, the results were either unsatisfactory or a total failure. The ironmasters found that either the metal was what was known as 'cold short', that is, brittle at low temperatures, or 'rotten', over-oxidised.

The reason soon became apparent. Bessemer had been lucky. The iron he had used in his experiments had come from Blaenavon, where the pig iron was virtually free of phosphorus. At Ebbw Vale and Dowlais, the pig iron, like most British pig iron, was high in phosphorus, and the lining they used for their furnaces was acidic, which meant that it failed to react with the metal after it too had formed an acidic compound during processing. The result was that the phosphorus finished up in the steel from the converter, which caused the cold short condition.

The manager of the Ebbw Vale Company took a sample of the Bessemer metal to a leading metallurgist, Robert F. Mushet, who lived in the Forest of Dean. He recognised the over-oxidation effect and re-melted it with the addition of spiegeleisen, a compound of iron, carbon and magnesium, which increased the carbon content and corrected the fault. Bessemer tried to get Mushet to explain what he had done, but the metallurgist explained that he was bound to respect the Ebbw Vale agreement not to make his findings more widely known.

A number of foreign companies also tried the Bessemer method, and it was Göran Göransson, the manager of a Swedish company, who actually made a major improvement in the method. Bessemer had originally had to re-melt the metal before casting it, but Göransson, by improving the airflow, was able to cast direct into ingots. The system could be made to work satisfactorily, providing iron made from phosphorus-free haematite ore was used.

Bessemer decided that the best way to make money from his invention was to set up his own steelworks in Sheffield in 1858. Meanwhile, he had improved on his own work. Instead of the ordinary, cylindrical furnace used for his initial experiments, his new converter was egg-shaped with a spout set at an angle to the vertical, and mounted on axes so that it could be tilted down to a horizontal position. To start the process, it was tilted so that the spout pointed upwards, to allow the metal to be poured straight in without losing any of its heat. It was then turned back to the vertical position until the process was complete. At the end it was tilted down in the opposite direction to the one used at the start, with the spout pointing down, to allow the steel to flow out.

Bessemer made a fortune from his invention. By 1870 he had earned at least £1 million in royalties from around the world, around £50 million at today's value. It did not stop him from continuing in the world of invention, although not always with the same degree of success.

He designed a new type of ship, which he named after himself. The *Bessemer* was a cross-Channel ferry launched in 1875, and its designer claimed that it would prevent seasickness. The hull and machinery were conventional, but the passenger section was suspended inside it. Hydraulic gear was supposed to ensure that, no matter how much the ship might pitch and roll, the central section would remain on an even keel. Sadly, Bessemer had not allowed for reaction time: the hydraulic mechanism was constantly trying, and failing, to keep up with the motion of the hull. The effect, far from calming the digestive systems of the passengers, only succeeded in making them even queasier.

The *Bessemer* may have been a failure, but Bessemer steel was to become an essential element in ship construction. By the time he died in 1898, Bessemer had received honours from learned societies around the world and been knighted. He also had the satisfaction of seeing his invention still in use, and it was to remain in use for nearly another century. The last Bessemer furnaces in Britain only ceased work in the 1970s. The author was fortunate enough to visit a steel works to see them in operation, and it was a sight never to be forgotten. A converter in action, with its shooting flames and flying sparks, must be the most spectacular sight that any industry has ever produced.

Bessemer steel did have an impact, but it was clearly not the answer to every problem that faced those who wanted to make steel on a massive scale. The next major breakthrough was the work of two German brothers, William and Frederick Siemens. Their father, Christian Siemens, was a farm manager at Lenthe in Hanover, who had a large family of eight sons and a daughter who lived to adulthood. He died in 1840 at the age of just 53, and responsibility for bringing up the younger members fell on Werner, the eldest brother, who was then an officer in the Prussian army. William, who was 17 when his father died, later took over the responsibility of looking after the youngsters.

William was originally trained as an engineer, specialising in developing new ways of using electricity. He and his brother Werner devised new systems of electroplating, and it was decided that the place to develop the idea was England. William was sent over, with only a rudimentary understanding of the language. He was looking for someone to undertake the manufacturing process, so he assumed logically, but comically incorrectly, that the right person to go to was an undertaker!

The Siemens brothers became involved in all kinds of new inventions, including the regenerative steam engine and condenser. Going right back to the early days of the steam engine, a key element had been James Watt's separate condenser, in which steam from the cylinder was condensed in a separate vessel, creating a vacuum. It worked well, but it involved wasting a lot of heat energy. The Siemens' idea was to pass the exhaust steam from the engine through a heat exchanger. The partially cooled steam was then condensed, and the water from the condenser was then passed back through the heat exchanger, to be returned to the boiler at a high temperature. It was an ingenious idea and the Birmingham engineering company Fox & Henderson paid William Siemens a substantial sum of money for the patent. It enabled William to devote his time to new inventions, one of which, a water meter, was to provide him with a good income for years to come.

William began to investigate other uses of the idea of the regenerative process, working with another of his brothers, Frederick. The most important application was in the regenerative furnace. In traditional furnaces, the hot waste gases simply went straight up the chimney. Frederick took out a patent in 1856 that found a way of using the heat. The gases were passed through a chamber lined with refractory bricks. These heated up, and when they were hot enough, the exhaust gases were turned off, and air for the furnace blast was passed through the chamber. This meant that the blast was hot when it reached the furnace, providing a great saving in fuel. In effect, the heat generated in the furnace was reused to heat the furnace again. The process was taken a stage further by using coal gas instead of solid fuel at the start of the process. The idea was applied to all kinds of industrial processes where high temperatures were needed, with glass making among the first to use the system. But it was the application to steel manufacture that was by far the most important.

The Siemens brothers were not having a great deal of success in persuading British steel makers to take up their idea, though it was being used in iron making. But two Frenchmen, Emile and Pierre Martin, set up a furnace which was charged with a mixture of pig iron and scrap metal. As it was the first to be used for steel making, the process became known as the 'Siemens-Martin method' – when a system was later developed for making steel direct from iron ore, that process was distinguished from the other by being known simply as the 'Siemens process'.

William Siemens was still unable to interest British manufacturers, so he set up his own works in Birmingham in 1866. He called it the Sample Steelworks, and once he had achieved success he took out a patent for what was called the 'open hearth process'. A year later production got under way at a new factory, the Landore Steel Company Works in Swansea. Production at first was a comparatively modest 75 tons a week, but now the process could be seen to work, the open hearth furnace began to spread.

The system consisted of a hearth that held the molten metal, usually pig iron, scrap iron or a mixture of the two. The furnace was fired by the coal gas, alternately from either end. The waste gases then passed out into one of the two regeneration chambers, which contained a chequerwork pattern of bricks that absorbed most of the heat. The gases then finally escaped through the chimney. The regeneration chambers were used alternately, so that as one was heating up, the other was supplying heat to the process.

A flux would be added to the iron that would form slag to carry away unwanted elements. At the end of this period, the process of refining began, removing just the right amount of carbon by oxidation. The open hearth process was very much slower than the Bessemer, but this turned out to be an advantage rather than a problem. During the long refining process it was possible to take samples from the molten steel to discover exactly when the correct required amount of carbon remained. The process could then be halted, and the steel run off. With the Bessemer process, once things were under way there was nothing to do but let it take its course. The two processes complemented rather than challenged each other.

The Siemens brothers were to go on to become famous, not only for their contribution to steel manufacture, but also in the field in which they had started their careers – electrical engineering. They were among the pioneers who developed the electric telegraph and had a vital role in the design and manufacture of electric generators. It was a Siemens generator that was the very first to supply power for electric lights for ordinary houses at Godalming in 1881. William Siemens was undoubtedly one of the inventive geniuses of the nineteenth century.

The outstanding problem with the Bessemer process remained that of phosphorus. At some time in 1870 George Chaloner, a lecturer in chemistry at the Birkbeck Institution, now part of London University, told his class that 'the man who eliminated phosphorus by means of the Bessemer converter will make his fortune'. In the class that day was a young man called Sidney Gilchrist Thomas. He decided there and then that he would be the man to find the solution and make that fortune.

Thomas was born in Islington in London in 1850, the son of a clerk at the Inland Revenue. He was educated at Dulwich College, where he showed an early enthusiasm for science. His chances of going on to university disappeared when his father died while Thomas was just 17. He had no option other than to find work and managed to get a job as a junior clerk in the police courts. In his spare time he continued his studies, and it was then that he heard the lecturer talk about Bessemer. He now began doing as much research as he could with limited resources, all the time continuing his studies and taking exams. He was even offered a job as an analytical chemist by a brewery, but as the grandson of a Nonconformist minister and ardent teetotaller he felt bound to turn down the opportunity. He became a self-made expert on the subject of iron and was soon contributing papers to the technical journals.

He eventually came to the conclusion that the solution to the problem was to change the lining of the converter, replacing the existing material with one that was a chemical base and hard wearing. Existing furnace linings were acidic, and as any phosphorus in the metal would have been turned into phosphoric acid, there was no way in which the acidic materials could react. But a base lining would react with the phosphorus. He reached his theoretical conclusion in 1875, but it was not until 1877 that he was able to experiment at an actual ironworks. Fortunately his cousin, Percy Gilchrist, was a chemist at Blaenavon and he agreed to help set up trials.

By 1877 Thomas was able to take out his first patent, and by 1878 the results were so promising that he was finally able to resign from his job at the court and devote himself full-time to his work on the converter. By 1879 he had the answer, using a form of dolomitic limestone for the furnace lining and lime in the early stage of the blow to form a slag. Not only had he improved the basic Bessemer method, but the phosphorus-rich slag proved to be a valuable side product, used in farming. The new technology was used under licence by a steel manufacturer in Middlesbrough in 1879.

Thomas had the success for which he had worked so hard, but did not have long to enjoy his success. His long years of working full-time in the courts followed by hours of study at night and at weekends had taken their toll. He tried to recover his health by travelling round the world, but the hoped for improvement never materialised. He died of emphysema in Paris in 1885.

Not all Bessemer converters were altered – where good phosphorus-free pig iron was available there was no need – but at least the options were all available, and by the end of the century steel output had soared. In 1896, 1,845,000 tons of steel was produced by Bessemer furnaces of both types and 2,355,000 tons by Siemens furnaces. With steel available in such huge quantities around the world it was inevitable that soon new uses would be found and that the Age of Steel would indeed be quite different from the Age of Iron.

12

THE TRIUMPH OF STEEL

Once a new material becomes readily available, people soon find new uses for it. After the initial setbacks, Bessemer steel became more generally accepted.

One very significant development took place on the Midland Railway. There was a particular section of track at Derby that was so heavily used that the wrought-iron rails had to be replaced every three months. It was decided to try the new steel rails on one length of track as an experiment. They were still in place fifteen years later, in spite of having 500 trains a day thundering across them. By 1864 the Midland directors had been totally convinced of the superiority of the new material and established their own plant for making Bessemer steel at its works at Crewe.

The steel rail was to become standard throughout the railway systems of the world, and it helped one man make an unprecedented fortune. Andrew Carnegie's story is often taken as the perfect example of a rise from rags to incredible riches. He was born into the family of a weaver in Dunfermline in 1835. Surprisingly, perhaps, at such a late date, his father was still working in the one-room cottage on a handloom. It was a doomed occupation as the unstoppable tide of industrialisation swept over all the textile districts of Britain, and the far more efficient power looms installed in mills reduced the workers in the old cottage industry to penury.

In 1848, the family were able to borrow enough money to emigrate to America, settling in Allegheny, Pennsylvania. The city was chosen because it was already a centre for the cotton industry, with a number of mills employing thousands of workers. When the family arrived, the industry had only recently recovered from a six-month strike which, although the workers failed to gain the decent wages they hoped for, did at least highlight the

Carnegie's steel works at Homestead, Pennsylvania, drawn by Joseph Pennell.

appalling conditions in many of the factories. As a result an Act had just been passed limiting work in the mills to ten hours a day.

It was here that the 13-year-old Carnegie started work as a doffer, removing the bobbins as they filled with yarn and replacing them with empty ones. It was a job with few, if any, prospects and he soon left to become a telegraph messenger boy at double the pittance he had received in the mill. He showed immediate aptitude and made a special point of learning about the machinery as well as delivering messages, so that within a year he was made a telegraph operator. All the time he was continuing to educate himself, borrowing books from the free library, made available to working boys by a local beneficiary.

Carnegie was a young man who combined many talents, not the least of which was a desire to learn from everything he did. It was not long before he was on the move, this time to take a small part in one of the most important and dynamic industries of the day. He joined a railroad company. In 1853, at the age of 18, he was given a job as secretary to Thomas A. Scott, the superintendent and financial director of the Pennsylvania Railroad Company.

He rose rapidly within the company and learned a lot from Scott, though he was not perhaps the most moral of teachers.

The railway barons of that age were notorious for corruption. Scott was later to be involved in the Union Pacific, a railway that was to form a vital link in the first coast-to-coast line. A typical example of the chicanery involved happened right at the start of construction. The government was offering a land grant alongside the track for every mile built, and the engineers were instructed to lay out the first section out of Omaha, not on the obvious direct line but on a circuitous route that added 9 extra miles to its length. It was nothing to do with engineering, but it took the route over land that would be needed as Omaha expanded, and which could be sold on at a huge profit. It is interesting that Carnegie, a man noted for his probity and philanthropy, learned a lot about business from a man noted for neither. As it was, Carnegie continued his advancement through the company.

These early years have been dealt with in some detail, because they are indicative of just how talented and industrious the young man was – to rise from a doffer at $1.20 a week to become superintendent of the Pittsburgh Division of the railroad company within just a few years. There was one vital lesson he learned from Scott: to make money, you need to invest. He was encouraged to put $500 in Adams Express, a company founded in 1854, and Scott knew the investment was safe, as Adams had already agreed to take over the messenger service on the Pittsburgh Railroad. Basically, this was insider dealing and therefore illegal. Carnegie's mother was persuaded to take out a mortgage on their house to provide the capital. It proved to be just the first of many excellent investments. Young Carnegie was starting to build up a useful capital for himself. Progress was interrupted by the Civil War of 1861–1865, but after the war he was back making money again.

After the war years, he made a number of trips to Britain, selling bonds in American railroads, and it was here he first heard about the new Bessemer process. Back in America, one of his investments proved sensationally profitable. He put $40,000 into Story Farm, Oil Creek, Pennsylvania. The name was not accidental – the oil was soon flowing, and dividends of $1 million a year were being paid.

He broadened his interest to take in ironworks and eventually formed two companies, the Keystone Bridge Works and the Union Ironworks in Pittsburgh. He worked with an engineer, James Eads, to design and build a bridge across the Mississippi at St Louis. They were, however, dubious about the advisability of using iron in the construction. He was very wise, showing more care than a British engineer who was also working on an iron bridge for another wide river crossing. Whether Carnegie actually took much notice of the events on the other side of the Atlantic or not, events in Scotland were to become a turning point in the move from construction in iron to building in steel.

Work on a bridge across the 2-mile-wide River Tay at Dundee began in 1871 under the direction of the engineer Thomas Bouch. It was the longest bridge ever built at that time and consisted of fifty piers joined by iron lattice girders. On a stormy night in 1879, just one year after it had opened, the bridge collapsed, hurling a passenger train into the water. Altogether, seventy-five lives were lost. Bouch was in disgrace, and his plans for another long railway bridge, across the Firth of Forth, were scrapped. A new design was produced, using steel instead of iron, and is perhaps the most famous railway bridge in Britain and happily still stands today.

Turning back to the Carnegie story, he was recognising the fact that steel was the material of the future and he could make his fortune by making steel himself using the Bessemer process. In the event, he decided that, not only could the Eads Bridge be built, but it could be built using Carnegie steel. It remains an impressive structure, over 6,000ft long (1.9km) with its widest span an impressive 520ft. It was revolutionary for its time, and at the opening a circus elephant was led across to demonstrate that it really was safe enough to carry heavy loads. This was unnecessary, because the real tests had been carried out long before that.

One of the reasons for the Tay Bridge collapse was that the structural members were not strong enough to stand the strain of the gale. No one could have known that at the time because there was no means of testing materials with any accuracy before construction began. This was no longer the case when James Eads came to design his bridge, thanks to the work of a Scotsman, David Kirkaldy.

Kirkaldy was born in Dundee in 1820 and went to work at the famous engineering works of Napier's Vulcan Foundry. In the 1850s he began work on designing machines to test the strength of materials, took out a patent and in 1865 resigned from Napier's to develop his invention. By 1874 he had set up business in Southwark Street, London. At the centrepiece of the works was his main test rig. It was hydraulically powered, and the material to be tested was secured in the rig. It could then be stretched to test its strength under tension, or pushed to test for compression. It was rated as working up to a pressure of 1 million pounds, 6,700psi.

It was here that the parts for the Eads Bridge were sent for testing. It marked a real advance in the safety of structures. What engineers now had was summed up in the motto still to be seen carved above the door: 'Facts not Opinions.' The Kirkaldy Works are still much as he left them, with all the original machinery in place and in full working order. It is a site of the greatest historical and international importance, but at the time of writing is under threat of closure. There are encouraging signs, however, that it will be saved and will enjoy a secure future.

The use of steel transformed bridge construction, but an even more important use for the new material, as far as Carnegie was concerned, was in providing steel rails for the rapidly expanding American network. This task was made that much easier by his long association with Scott and his railway empire. By the 1880s Carnegie was the biggest supplier of steel rails in the world, and to start the steel making process his furnaces were turning out 2,000 tons of pig iron a day.

It was during this period that he bought out his biggest competitor, Homestead Steel Works. Carnegie has always been seen as a good employer, and a man who thought it unreasonable to take too much money for himself. He reckoned that $50,000 a year was enough – not exactly a small amount, but one that would be considered hopelessly inadequate by many of today's billionaires. He famously wrote that the man who died rich 'died disgraced'. He is, of course, one of the most famous philanthropists of all time – when he died, far from disgraced, he had given away nearly $400 million – roughly $5 billion at today's prices. He was in many ways an admirable man, but Homestead was to provide a major blot on his character sheet, even though the events happened while he was out of the country.

Carnegie had always claimed to be sympathetic to trade unions and the process of negotiation. The same could not be said of his manager at Homestead, Henry C. Frick. The industry was suffering a fall in the price of steel, which Frick intended to balance out by cutting wages. At the same time, he was quite prepared to ferment a quarrel in the hope of ultimately breaking the powerful Iron & Steel Workers' Union. In 1892 Carnegie was away on holiday in Scotland, but he left clear instructions to Frick that, if the union would not agree to terms, he should close the plant and lock out the 3,000 workers. He was unambiguous, writing to Frick that he would 'approve of anything you do', and would be with him to the end. It was the carte blanche Frick needed.

Over 1,000 men were locked out, and were told that they could negotiate to get their jobs back, but only if they agreed to leave the union. Although only a proportion were actually in the union, there was a virtually unanimous vote to stay out. Frick now turned the plant into a fortress and tried to hire deputies to guard it for him. They were run out of town. He then did what many other industrialists were doing at the time, called in the Pinkerton Agency to help. They had a private army, with a reputation as strike breakers – and of not being very particular about the methods they used in the process.

On 5 July tugboats appeared on the river, hauling barges filled with armed Pinkerton men. The alarm was spread and a crowd gathered, telling the Pinkerton men they would not be allowed to land. At some stage in the proceedings a shot was fired, though it was never clear which side fired first. After that a tremendous fight developed, with the crowd on shore using everything from firearms to throwing sticks of dynamite at the boats. After a long battle, the Pinkerton men gave up their attempt to take over the works, but by then three of their men and nine workers had been killed.

It was a short-lived victory for the workers. The Governor of Pennsylvania now stepped in and ordered the state militia to the plant, armed with modern rifles and machine guns. It was an act of dubious legality, to use state troopers to intervene in a purely industrial dispute between management and men, especially when it was the company that had locked out the workers. Under the militia's protection, new workers were brought in from outside. Resistance crumbled, and ultimately Frick had achieved his objectives. He had established control over wages on his own terms and effectively broken the power of the union.

Throughout all these proceedings, Carnegie remained aloof and detached, refusing to become involved. He was holidaying in Scotland and refused to receive any communications about the business from anyone except Frick. There was more than a whiff of the Pontius Pilate about his attitude. He would later declare that the affair was 'the trial of his life'. In a letter to William Gladstone, he wrote that 'nothing before or since wounded me so deeply', and 'the works are not worth one drop of human blood. I wish they had sunk.' Of course, the works did not sink, but were continued with non-union labour, in spite of Carnegie's protestations of distress. There was a failed assassination attempt on Frick. The men, it seems, still believed that they would get a fair deal if only Carnegie himself was the man to appeal to, with Frick out of the way.

In 1901, Carnegie was 66 years old and ready to retire. He was bought out by the powerful banker and financier John Pierpont Morgan, and his works formed the basis for the giant United States Steel Corporation. Nothing could have signified the importance that steel had achieved more completely than this great enterprise, with a capital value of $1 billion. Meanwhile Carnegie, who had built his fortune on that very first investment using $500 borrowed from his mother, set about getting rid of as much of his fortune as he could in his many charitable bequests. Among those who benefited were public libraries in Britain and America – he had never forgotten how much he had been helped as a boy when he had been able to borrow books for nothing.

Bridge building was revolutionised by the ready availability of steel. The Forth Bridge was Britain's first bridge to be built from mild steel, and it is worth looking at just what was entailed in building such a structure. The men responsible for its construction were the engineers John Fowler and Benjamin Baker and the contractor, William Arrol. All three men were to receive knighthoods.

Their design is commonly known as a 'cantilever' bridge, but it actually consists of vast diamond trusses joined by girders. The engineers themselves referred to it as a 'continuous girder' bridge. A key element in the siting of the bridge was the little island of Inchgarvie in the middle of the Firth. It was to be the base for the central support, but even then it was too small to take the whole massive structure, so the contractors had to find a way of reaching down to a firm foundation on the seabed. Other sections had to be built offshore as well.

Although it is often spoken of as one of the triumphs of British engineers, the early stages relied on workers from overseas. In order to get down to bedrock, the contractors had to construct caissons. These were giant cylinders 70ft across and as much as 90ft high. They would be sunk to the seabed, and compressed air would be used to keep them watertight. This was a technique that had already been used on a number of projects on the Continent, so the contract went to an Italian company who brought over French workers.

It proved to be a difficult operation. The lower rim of the caisson was sharpened to cut through mud and silt, which turned out to be much deeper than anticipated, and at the Queensferry end the bottom of the largest caisson was 89ft below the high water mark. Working within the caissons was difficult, uncomfortable and dangerous. Many workers suffered from what was known as 'caisson disease', very similar to the 'bends' experienced by deep-sea divers. It was caused by coming up too suddenly from the high pressure of the compressed air chamber. Human blood contains dissolved gases, notably nitrogen. As the pressure lowers it is possible for these gases to escape, creating bubbles in the blood vessels.

The men who did the actual construction work mainly came from the Scottish engineering industry and were known as 'briggers'. It was a gargantuan task. Altogether, the bridge contained 54,000 tons of steel and the plates were held together by approximately 7 million rivets. At the height of construction there were 4,000 men at work, and before it was completed fifty-seven had died in accidents. The authorities, inevitably, put the accidents down to recklessness. Certainly there were none of the safety measures in place that one would find on a comparable engineering project today. The men wandered around the high scaffolding and over the girders as if they were strolling down a pavement, and there was a certain amount of trouble over drunkenness. Hard hats were unknown, so that a dropped rivet would be enough to kill a man. But at the end, the official report described it as 'a wonderful example of thoroughly good workmanship'.

It was not just on the railways that steel came into its own for bridge construction. A new material was available that enabled a different type of suspension bridge to be built. Making iron into wire was practised long before the start of the railway age, and the basic technique remained the same – drawing the hot metal through a series of dies. There was, however, a major new development when Joseph Horsfall invented the system of 'patenting'. He produced high tensile steel wire in a continuous process. A single strand was passed through the furnace, then through a quenching bath and finally through a bath of molten lead to temper it. The steel wire could be twisted without losing any of its strength, so strands could be used to form ropes and cables. Brooklyn Bridge, in New York, is the most imposing early example of using steel wire for a suspension bridge.

The story of the Brooklyn Bridge is one beset by tragedy. The original design was prepared by John Roebling, an experienced civil engineer and bridge builder, who had emigrated to America from Germany. While he was surveying the site he was involved in a serious accident, when his leg was crushed between the hull of a ferryboat and a piling. The damaged toes were amputated, but hygiene was not particularly strong in those days and he developed tetanus and died shortly after. His 32-year-old son had been given a suitable name to reinforce the family's new status as American citizens, Washington Roebling. He took over the management of the construction when work began in 1870.

As with the Forth Bridge, caissons had to be used to get down to foundations on the river bed and again, as with the Forth Bridge, many of the workers suffered from caisson disease. Among those who suffered was the engineer himself. It left him so weakened that he was unable to get to the site, and could only watch the work from his window. But he was determined to keep the project in the family and to maintain his own control, even if he could not be physically present. He needed someone he could trust to act as a go-between, to consult with him and to oversee the work on site. He decided that the ideal person for the job was his wife, Emily.

Men at work on Brooklyn Bridge.

It must have seemed an extraordinary idea to have a woman working in a managerial role on a major engineering project at that time. It would be newsworthy today in most places, and it must have been startling in the nineteenth century. Emily had to do far more than just pass on instructions. She had to be taught the mathematics needed to understand the physical principles of suspension bridge construction, the geometry of curves, the calculation of stresses, and be familiar with all the different materials to be used, from the granite blocks of the towers to the tensile strength of the steel cables. She and her husband remained in charge for eleven years. Many at the time believed that, in the latter years, it was Emily who was the real force behind the whole project.

At the official opening on 24 May 1883, it was Emily who rode across the bridge beside the President Chester Arthur. Her husband had to make do with a banquet at their home. Among those present on that day was one of Roebling's competitors, Abram Hewitt, who wrote a very fitting tribute to the woman who, perhaps more than anyone, had made the project a success. The bridge, he wrote, was 'an everlasting monument to the self-sacrificing devotion of a woman and of her capacity for that further education from which she had been too long disbarred'.

Emily was to go on to prove that her capacity for further education was indeed real. At the age of 56 she received a law degree from New York University. But it is the Brooklyn Bridge that remains the greatest tribute to a truly remarkable woman.

Steel had come to replace iron in all kinds of different ways, from the hulls of ships to the cables for suspension bridges, and to such an extent that wrought iron production soon more or less ceased. Mild steel took its place. The availability of large quantities of steel caused profound changes in transport, from the increased efficiency of the railways to a new method of constructing ocean liners, and in the process had made some men into multi-millionaires. But, as the nineteenth century drew to its close, a new use was found that was to change the world we live in, and whose effects are still being felt today.

13

ONWARDS AND UPWARDS

The structural use of iron has a long history, and the idea of using it for tall buildings was not new. The most spectacular example of a wrought-iron construction of the nineteenth century has to be the Eiffel Tower, but it is something of an anomaly. When it was completed in 1889, steel was already replacing iron in many structures, such as the Forth Bridge.

The next logical use would be to replace iron with steel as a framework for buildings. The fireproof mills of the Industrial Revolution had set a pattern, and it was clear that with a suitable steel frame it was possible to build multi-storey buildings. There was, however, a human limit to the height that they could reach – nothing to do with the limited imagination of engineers, but decided by how many stairs the building's users would be prepared to climb. (When the author's daughter and her family moved to a fourth-floor flat in Berlin, this parent felt that, with ninety-five steps to climb, that was quite high enough – though there was a further floor above theirs.) The answer to the problem of how to get to the top of high buildings without collapsing from exhaustion was resolved by the American engineer Elisha Graves Otis.

Elevators for carrying both goods and passengers had been in use in America since the middle of the nineteenth century. They were worked hydraulically and operated using a hydraulic ram. To increase the speed, the hydraulic system had its power amplified through pulleys, but now the elevator was suspended from a rope. Ropes fray and break, and if that happened with a freight elevator then no doubt it could cause a lot of damage, but that could be covered by insurance. It was a very different matter if it gave way while carrying passengers.

Otis demonstrating his safety elevator in 1854.

Otis was an unlikely candidate for the job of solving this problem. Born in 1811, he started his working life driving wagons, but later became a skilled mechanic and was taken on by the New York bed manufacturer Maize & Burns. The company needed a hoist to move heavy equipment to the upper floors of the factory, and it was Maize who encouraged Otis to work on a safe hoist to do the job.

So it was that Otis invented a safety lift originally intended for iron bedsteads, but which proved perfectly safe for people as well. There was a ratchet on each side of the lift shaft. The prawls were set on the sides of the lift, but in normal use when the rope was in tension they ran free. If, however, the suspension rope broke and the tension was lost, the prawls were immediately activated by springs to engage with the ratchets, bringing the whole system to a halt. He exhibited his new device at New York's Crystal Palace Exhibition in 1854 and gave a dramatic demonstration of its effectiveness. An elevator was installed, and Otis stood on top of it. As soon as a crowd had gathered, he cut the rope with an axe. It was a brilliant piece of showmanship and a dramatic demonstration. Hydraulic lifts could now be used with complete safety.

In 1883, the London Hydraulic Power Company was set up to provide, among other things, power for lifts in high buildings. It provided a lot of business for ironwork manufacturers. A series of power stations were set up, in which water was pumped up into tall accumulator towers and then released through narrow cast-iron pipes spread throughout the city. Without a suitable power source and a guaranteed safety regime, tall buildings would have been possible to build but impossible to use.

It was in America, not London, that the tall building was to come into its own, buildings so imposing that they were given a new name – 'skyscrapers'. It could be said that the development of the skyscraper owed a lot to Mrs O'Leary's cow …

Pennell's drawing of a skyscraper under construction: The New House, Philadelphia.

On 8 October 1871, a fire started in the barn behind the Chicago house of Patrick and Catherine O'Leary, popularly said to have been caused by the cow kicking over a paraffin lamp. The fire service were called, but went at first to the wrong address, and by the time they reached the O'Leary's it was well ablaze. It had been a very dry period and the surrounding houses and commercial buildings were all built of wood. The firemen did their best, but failed to contain the blaze that raged on for two whole days, destroying an area of more than 3 square miles in the heart of the city.

When a reckoning was finally made, it was estimated that some 300 people had died, 100,000 homes had been destroyed and almost $2 million worth of property had been lost. Rebuilding was essential, for Chicago was a rapidly growing city and real estate had become incredibly valuable. Now there was an incentive to build high, to make as much money as possible out of each plot of land that the devastation had made available.

The man who first met the challenge was William Le Baron Jenney. Jenney was born in Fairhaven, Massachusetts in 1831, into a wealthy family who made their money in whaling. Not unlike many young people today, after finishing in school he went travelling, going round South America and across the Pacific to Hawaii and the Philippines. It was there that he saw houses constructed on a bamboo framework, a method that ensured both strength and enough flexibility to withstand the typhoons of the tropics. Although there is no proof, it could have influenced his own ideas on construction.

In 1850 he went to Harvard to study engineering, but found the courses there to be too far removed from the practical world. He heard that things were very different in France, and he enrolled at the École Centrale des Arts et Manufacture in Paris, an institute that had also taught the young Gustav Eiffel. He graduated with honours and had hardly had time to take up his new career as a structural engineer, with a short period working with a railroad in Mexico, before America was swept into Civil War.

He returned home to enlist in the Union Army Corps of Engineers, ending the war with the rank of major. In 1867, he moved to Chicago to set up in business, not as an engineer, but as an architect. It was a rather surprising shift in direction, but it was fortuitous as it was his engineering skills that would prove vital to the new development.

The fire of 1871 had inevitable consequences. There was an urgent need to rebuild, and clearly the new buildings would have to be more resistant to fire than the old. Chicago was booming, and by the middle of the nineteenth century it had become the transport hub of America, with fifty lines entering the city. Businesses thrived and everyone wanted offices near the centre. Land prices were high and the best way to capitalise on that was to make the best use possible of each patch of land by building tall.

High buildings were not new, but in order to support the upper floors the lower walls had to be immensely thick. One of the last buildings to be constructed on the old principles was the Pulitzer Building in New York, completed in 1890. It was fourteen storeys high, but the masonry walls at the base were 9ft thick. This was not just expensive but also took up a lot of valuable space, and in order to maintain the structural integrity, windows had to be kept quite small. Jenney was to pioneer a very different construction technique.

In 1878, Jenney was asked to design a department store for Levi Z. Leiter. It was to be five storeys high – later extended to seven. He decided to use an iron skeleton as the main structural frame, with terracotta fireproof material covering all the various load-bearing structures. Because it was on a corner site adjoining other buildings, the city authorities insisted that the party walls be masonry – they clearly had little confidence in the new ideas. The rest of the building had a thoroughly modern appearance, with iron pilasters separating the large window spaces. It was not yet quite a skyscraper, but it had set a pattern.

In 1883, Jenney set to work on an even more ambitious building for the Home Insurance Company. This was a ten-storey office block and was genuinely revolutionary. For the six lower floors, he used a similar technique to that used in the Leiter Building, with cast-iron pillars encased in brickwork and wrought-iron beams to support the floors. Above that, the remaining four floors were supported on beams of Bessemer steel. Now the modern skyscraper had arrived and steel had made its first appearance in this type of structure.

Inevitably it seems, popular mythology has to have some anecdote to show how Jenney hit upon his new approach. The story is that he came home from work one day and found his wife reading. As she got up to greet him she put her heavy book down on top of a wire birdcage. Jenney was impressed: he kept picking up the book and dropping it back onto the cage. If flimsy wire could support a heavy book, then perhaps an iron frame could hold up an entire building. It was his eureka moment. The story might even be true, but it seems far more likely that his success lay in the fact that, unlike most architects of the day, he had initially qualified as a structural engineer.

Jenney had established what became known as the Chicago School of Architects. Structurally his buildings were new, but it was to be another architect who was responsible for rethinking the new style for the iron and steel frame building.

Louis Sullivan was born in Boston in 1856 and studied for a year at the Massachusetts Institute of Technology before continuing his architectural studies at the École des Beaux Arts. He formed a partnership with Dankar Adler, and together they designed the Transportation Building for the World's Columbian Exposition in Chicago in 1893. It was an immense structure and was notable for the fact that, unlike most of the other new buildings on site, it made no reference to classical architectural styles. He was scornful of what he saw as mere pastiche.

Sullivan was developing a new aesthetic for city buildings. In 1896 he wrote an essay, 'The Tall Office Building Artistically Considered', in which he wrote:

> It is the pervading law of all things organic, and inorganic, of all things physical and metaphysical, of all things human and all things super-human, of all true manifestations of the head, of the heart, of the soul, that the life is recognisable in its expression, that *form ever follows function*. This is the law.

That 'form follows function' was to be the mantra for the next generation of architects in the twentieth century. The emphasis was on the vertical rise of the building, not the horizontal. The facade was to reflect the internal grid-like structure.

In 1899, Sullivan, no longer working with Adler, designed the new Carson Pirie Scott Store, a twelve-storey building, the facade of which accurately reflected the structure. Thanks to its steel frame, it was possible to have large window spaces, ideal for showing off the wares to passers-by. It is notable that Sullivan had actually felt confident enough to propose an even more imposing thirteen-storey building.

The movement towards ever higher buildings was under way, and it was to reach its twentieth century apotheosis in a place where land values were at their highest: Manhattan, New York.

There is a famous photograph of men taking a lunch break while working on a New York skyscraper, sitting in a row on a beam suspended high over the city streets as nonchalantly as though they were having sandwiches at the bar of a diner. The most famous of all the groups who worked at these dizzying heights were the Mohawks. Their story was told by Joseph Mitchell in an essay in the book *Up in the Old Hotel*, published in 1949. They came from the Kahnawake reservation in Canada, where life had very little to offer, and they were attracted to the site of a nearby major construction project in 1886.

The Dominion Bridge Company (DBC) was building a bridge across the St Lawrence for the Canadian Pacific Railway. The company decided to employ a few of the local Mohawks as ordinary labourers. The Mohawks were altogether more ambitious. A DBC man told Mitchell exactly what happened next:

> They were dissatisfied with this arrangement and would come out on the bridge itself every chance they got. It was quite impossible to keep them off. As the work progressed, it became apparent to all concerned that these Indians were very odd in that they did not have any fear of heights. If not watched, they would climb up into the spans and walk around up there as cool and collected as the toughest of our riveters, most of whom at that period were old sailing-ship men especially picked for their experience in working aloft. These Indians were as agile as goats. They would walk a narrow beam high up in the air with nothing below them but the river, which is rough there and ugly to look down on, and it wouldn't mean any more to them than walking on the solid ground. They seem immune to the noise of the riveting which goes right through you and is often enough in itself to make newcomers to construction feel sick and dizzy. They were inquisitive about the riveting and were continually bothering our foremen by requesting that they be allowed to take a crack at it. This happens to be the most dangerous work in all construction, and the highest paid. Men who want to do it are rare and men who can do it are even rarer, and in good construction years there are sometimes not enough of them to go around. We decided it would be mutually advantageous to see what these Indians could do, so we picked out some and gave them a little training, and it turned out that putting riveting tools in their hands, was like putting ham with eggs.

The work was dangerous. During the construction of the bridge one of the beams collapsed, and thirty-five of the Mohawks died. It did not deter them. They realised that they had acquired skills that were in demand, and were soon making up gangs to travel south in the States to take advantage of the construction boom with its huge labour demands. The famous Empire State

Building employed over 3,000 workers at one time, of whom hundreds were the Canadian Mohawks who often took the most dangerous jobs.

The actual work of riveting was very much the same as that in other industries, such as ship building, except that the men were working high up in the air from simple wooden platforms suspended from the beams. On one of these simple platforms was the heating brazier, from which the red hot rivets were thrown to the gang and caught in a cup. No one seems to have given much thought to what might happen if one of them missed. That was the least of the dangers. As Mitchell rather nonchalantly remarked, one false step and it was 'Goodnight Charlie'.

The men who did the work were divided into experienced hands and newcomers. The latter were known as 'snakes' because they were considered deadly. They were the ones who, unused to the work, were most likely to take a false step or slip and then they would grab whatever was nearest to try and save themselves. Too often, the nearest object was a fellow worker. As one of the workers said, 'The thing I hate most is taking on a new man when we're near the top. They all get used to it or get killed.'

The popular view of the Mohawks as having some sort of inbuilt resistance to vertigo was denied by the men themselves, 'A lot of people think Mohawks aren't afraid of heights: that's not true. The difference is that we deal with it better.' One thing all employers seemed to agree on was that the men got restless, and if they decided the time had come to move on then they would be off, in spite of all arguments. George C. Lane of the Bethlehem Steel Company described a typical incident:

> In the summer of 1936 we finished a job here in the city and the very next day we were starting in on a job exactly three blocks away. I heard one of our foremen trying his best to persuade an Indian gang to go to the new job. They had heard about a new job in Hartford and wanted to go up there. The foreman told them the rates of pay were the same; there wouldn't be any more overtime there than here; their families were here; they'd have travelling expenses; they'd have to root around Hartford for lodgings. Oh, no; it was Hartford or nothing.

Lane met up with them a year or two later and asked how they had got on in Hartford. They never got there. They had ended up in San Francisco building the Golden Gate Bridge instead. The last word on these extraordinary workers should be those of one of their own, Orvis Diabo:

'I heated a million rivets. When they talk about the men who built this country, one of them is me.'

Iron and steel were to find a use in another form of construction for a variety of buildings. Concrete is not a new material, though we tend to think of it today as being quite modern, and in popular lore the word is almost synonymous with 'ugly'. Yet one of the most beautiful of all the buildings of the ancient world, the Pantheon in Rome, was built out of concrete. That it still stands today is astonishing, since no modern engineer would dream of constructing such an immense domed building out of concrete alone. He would specify reinforced concrete.

Concrete, rather like cast iron, is very strong in compression, but comparatively weak in tension, which makes it fine for use in load-bearing columns but less suitable for beams. As early as the eighteenth century, the architect J.G. Soufflot had attempted to strengthen conventional masonry by embedding iron rods, but there was a problem. The joints between stone blocks could let in water that would eventually corrode the iron and weaken the whole structure. The same problem does not arise if the rods are enclosed in concrete made from poured cement.

In the nineteenth century Portland cement had become a popular building material, and a French engineer, J.L. Lambot, made a material consisting of concrete containing an iron mesh. He demonstrated it at the Paris Exposition of 1855, not, bizarrely, in a conventional building but in a rowing boat. Although it attracted a great deal of interest he never developed the idea. That was left to another Frenchman, Joseph Monier, whose first efforts were similarly far removed from the construction industry. He started out making pots for ornamental orange trees.

Monier was a gardener, and he got tired of having terracotta pots that either broke through weathering or the growth of roots. So he built pots made of cement with a mesh of iron rods. They were successful and he realised that the process could be scaled up. In the 1870s, he began constructing reinforced concrete water tanks, the biggest of which at Bruyères à Sèvres had a capacity of 1,000 cubic metres.

In 1875 he designed a small concrete bridge and even built a demonstration concrete house. He was not quite the first with this application. Shortly after he had demonstrated his concrete pots, another French builder, François Coignet, created the first reinforced concrete house in Paris. It still stands today. Monier was a genuine pioneer, and invented what was to become a standard feature of many buildings – the concrete T-beam.

The end of the nineteenth century was a period of rapid development of the new material. Much of the early work took place in France, where François Hennebique developed a complete building system in which all the structural elements of a building could be made from reinforced concrete. In 1897, the Hennebique system was introduced into Britain, and it was a British engineer, E.L. Ransome, of the Ipswich iron manufacturers Ransomes & Sims, who took the idea to America.

In America, he discovered factories were still being built using the technology for fireproof mills that had been developed in Britain a century earlier, with brick arches springing from iron columns to support the floors. He introduced the idea of using tie beams embedded in concrete. He developed the idea into an improved system using pre-cast reinforced beams, and in 1903 erected America's first concrete-framed building at Greenburg, Pennsylvania. He used a new form of reinforcement, with twisted steel bars embedded in the concrete.

The idea was taken up by the architect Melville E. Ingalls, who proposed using it for a sixteen-storey building in Cincinnati, Ohio. It took him two years to convince the city authorities that the system was safe, but he eventually got his licence and in 1903 America got its first reinforced concrete skyscraper.

It had been an interesting passage, with developments being passed from innovator to innovator, and from country to country, but the world had a new building material that was to typify much of the architecture of the twentieth century. Iron and steel had won a prime role in the buildings of the world, whether used as a skeleton to be clad with a wide variety of different materials, or to be hidden from sight and yet having a crucial role in the integrity of the new concrete structures.

14

THE MODERN AGE

Modernisation in the twentieth century has become more or less synonymous with mechanisation. The iron and steel industry was heavily dependent on skilled men doing often dangerous, hot and uncomfortable work, and in the United States in particular there was a great shortage of hands with appropriate skills. With the demand for steel rising all the time, there was a need for all the processes to be speeded up, from the initial creation of the metal in blast furnaces through to working the finished product.

The first changes came in the foundries, not just to parts but to the whole system. America was installing ever larger furnaces that required more and more raw material. The first of the big, modern furnaces was installed at Carnegie's Edgar Thomson works in 1880. It was a monster, rising to a height of 80ft, with an 11ft-diameter hearth. Hot air blast was supplied by eight tuyeres fitted into the 20ft-diameter boshes. Furnace output rose from 800–900 tons a week to an impressive 1,200 tons. Soon, even bigger furnaces were being built, rising as high as 100ft and able to turn out 2,000 tons a week each. It was obvious that new technology was required to make the most of these new furnaces. The changes, quite literally, spread from top to bottom.

The top of the furnace is where the charge of ore, flux and fuel are added. In the past, the raw materials would be brought to the site and then wheeled up in barrows to the surface top or, on more sophisticated sites, carried up an incline or brought up in a mechanical lift, but it still had to be emptied into the furnace by hand. With furnaces producing up to 2,000 tons of iron a week, this involved a huge amount of raw material, perhaps as much as 5,000 tons a week, most of which would be made up of the heavy ore, together with coke and limestone. Mechanisation was the obvious solution.

A start had been made as early as 1849, when George Parry of Ebbw Vale began looking at ways of improving furnaces. The process began when ironmasters started to think of ways of using the gases that were produced during the firing. For years they had simply burned away at the top of the furnace, which was clearly wasteful. Various schemes were put forward for collecting the gas and then using it for heating the air blown into the furnace.

Parry invented a system in which the charge for the furnace was loaded into a hopper with a cast-iron bell beneath it. The bell was at one end of a counterbalanced arm so that when it was raised it closed off the hopper, as a result of which the gases from the furnace had only one means of escape, down a pipe where it was fed into the hot gas stoves. Charges were added to the hopper and when enough was in place, the bell was lowered, allowing the material to fall down into the mouth of the furnace. By means of this arrangement, the furnace top was closed off for most of the process, conserving the heat and putting the waste gas to good use.

The charge still needed to be distributed evenly throughout the furnace, a process that originally called for men with barrows to feed it in at different places along the furnace. An American invention took the process a stage further by having the charge loaded into a skip that was automatically emptied into a hopper, from which it was allowed to fall onto a rotating distributor to be passed to two separate bells. Using this technique, the whole process could be automated and the charge evenly distributed while keeping the top of the furnace permanently closed. As the top was never opened, no gas could escape and heat was conserved.

At the bottom of the furnace another important change was introduced. In the past, the molten metal had been run into moulds set in the sand of the casting house floor. The pigs generally weighed about a hundredweight each, so for an output of 2,000 tons a week that would mean the production would produce 40,000 individual pigs that all had to be taken from the moulds and wheeled away. In 1895, a new machine appeared in America that revolutionised the casting of pigs.

Now the moulds were mounted on a continuous belt that passed under a spout from which the molten iron was poured. As each mould was filled, it was moved on, cooled by a water spray and then the mould was turned upside down, dropping the pig into a wagon. It represented a huge saving in labour.

There was one other device that helped to optimise the output of the furnace. When a tap hole in the furnace had to be closed, the blast had to be stopped to allow everything to cool down to a temperature that allowed a man to get close enough to insert the clay plug. With the new technology, a device like a gun, powered at first by steam and later by compressed air, fired the plug into place. There was no longer any need to turn off the blast. Now furnaces could be run continuously, none of the valuable gas would be wasted and the pigs were supplied in a steady stream to the conveyor belt. The system saved on manpower, made the best possible use of the furnace and opened the way to increased production. Over the years, furnaces were to get even bigger and still more productive, which was just as well, as demand was still increasing.

It was not only at the start of the steel-making process that changes were being made. Throughout much of the nineteenth century, power for the rolling mills had come from increasingly powerful and immense steam engines. Now there was a new power source available – the electric motor. This had enormous advantages. It was much more versatile than the steam engine. It could be started and stopped at the flick of a switch, it was comparatively small and compact and you did not need to have one engine to run everything in the works – each individual piece of machinery could have its very own power source.

The steam engine was not immediately made redundant. In the days before there was such a thing as a national grid, if you wanted to use electric motors then you had to generate your own electricity. There were different ways of doing this. If there was a suitable water supply, then a water turbine could do the job, and with the arrival of the internal combustion engine, a diesel engine was available; but for most, steam was the first choice.

A company could choose how to generate its power, and an interesting survivor of those days, though not involved in the iron industry, is Longford's Mill near Nailsworth in Gloucestershire. This former woollen mill is no longer in use, but the power house has survived and shows how the company moved on from one device to another, with three sets of generators, one powered by a two-cylinder vertical steam engine, the second by a turbine fed by the large mill pond and the third by a diesel. If nothing else, it shows that industrialists in the early twentieth century were willing to try different ideas and make changes whenever a better alternative came along. Interestingly, it was the oldest form of power – the power of water – that remained the first choice.

Electricity reached the iron industry at some time in the 1880s, but was of comparatively minor importance, though the replacement of the old gas-lights with electric light must have been a bonus for everyone at work. But the early applications were only used for machines carrying out comparatively light work. In time, however, as bigger and more efficient electric motors became available it was to be the driving force behind everything used in the industry, including the new machines that were being introduced into the rolling mills.

In Britain, a new type of rolling mill, known as the 'universal' rolling mill, was first introduced to a steelworks in Middlesbrough in 1878. This had four rollers instead of the usual two, one pair being at right angles to the other. This meant that, by passing the sheet between the side rollers, it was possible to control the width of the resulting metal sheet with more accuracy than had ever been possible before. Henry Grey took the idea even further when he invented a roller that would produce beams instead of plates. When British manufacturers showed very little interest in his idea, he took his invention to Germany and the first mill went to work at Differdingen in 1902. The new machinery produced the H-beam that has become a standard form used on construction sites throughout the world today.

Germany was also the place that saw the first application of another, and very different, type of rolling mill. In the past, metal tubes had been made by bending sheet metal into a cylinder and welding the edges together. Two brothers, Max and Reinhard Mannesmann, were steel manufacturers at Remscheid. They developed a system for manufacturing seamless tubes. A hot metal rod is passed through rollers, set at an angle to each other and rotating in the same direction. The rollers pull the rod forward and a cavity begins to open up. The rod then passes on to a circular head piece that forms the cavity into a circular hole.

It might seem that this was not the most important development ever to hit the industry, but one great inventor at least saw its importance. When Thomas Edison visited the Chicago World Exhibition in 1893, he was asked which exhibit had made the most profound impression on him. His choice went to the Mannesmann steel tube. The simple steel tube was to have a vital role to play in two developing industries.

Metal tubes turned out to be just the thing for a new type of transport, the bicycle. First developed in France, it became a huge success there and, foreshadowing the now famous Tour de France, the world's first cycle race from Paris to Rouen took place in 1869. But the French were not the only ones to be enthused by the new machines. The race was won by an Englishman, James Moore.

That same year, Britain's first cycle manufacturer set up in business in Coventry. The 'ordinary', popularly known as the 'penny-farthing', dominated the early years. The rider sat perched above the enormous front wheel. It was so big simply because there were no gears, so the whole machine moved a good distance for each turn of the pedals, but it is not the easiest machine to use. It is not too bad once you are on board, but getting started is tricky – and getting off without toppling over even harder. Inevitably, manufacturers were always looking for new ideas and improvements.

James Starley was a Coventry manufacturer who brought out a tricycle with a chain drive from the pedals. At this time his nephew, John Starley, was being brought up in Essex, the son of a market gardener. The boy showed great aptitude for mathematics and science, and when he showed his uncle a drawing he had done of an oscillating steam engine, he was whisked away to Coventry to work in the family business. It was there, in 1885, that he brought out a new design – the Rover bicycle.

This is very recognisable as the model for all later machines, with two wheels of equal size, a chain drive from the pedals to the rear wheel and a triangular tubular frame. The demand for steel tubing increased as the popularity of the bicycle grew. It was a technology that was also to prove vital for the development of another new form of transport.

Over in America, two brothers ran their own bicycle repair shop, so they were used to the idea that if you wanted something with a frame that combined light weight and strength, then the metal tube was the obvious answer. And that is just what they needed for their ambitious project. They wanted to build a flying machine. The brothers were, of course, Orville and Wilbur Wright.

They also had a suitable power source that was proving itself on the road, the petrol engine. They made their first flight on 17 December 1903. It was not exactly spectacular, lasting a mere twelve seconds. Unlike most later aircraft, this had a pusher propeller mounted behind the pilot, and in order to reach airspeed it had to be launched down a ramp. There was no tail and in order to bank when changing direction, the pilot had to shift his body to one side or the other. It is not easy to manage. Anyone visiting the excellent Swiss Museum of Transport in Lucerne can try for themselves on a simulator. It is good fun and, unlike the real thing, quite painless if you crash it.

So it was that a simple invention for turning a metal rod into a hollow tube proved to be essential in the development of two important, but very different, forms of transport.

It was not just in the methods of production that there were huge and important changes as the nineteenth century drew to a close. Just as important were the changes in the basic materials themselves. The man who led this revolution in steel manufacture was Robert (later Sir) Hadfield. His family had many connections in the local iron, steel and cutlery industries and his father set up the Hecla steelworks in Attercliffe, near Sheffield, concentrating on what was then a very new technology – casting in steel.

Robert was born in 1858, went to school in Sheffield and proved to be a good scholar. He could have gone on to university, but chose instead to enter the family business. He entered the industry at a critical time. Where earlier ironmasters had worked on a more or less pragmatic basis of trial and error, there was an increasing awareness that the newly developing sciences would be crucial to future developments. Robert devoted a great deal of time to studying chemistry and metallurgy and began applying his knowledge to the development of steel alloys. He conducted many experiments using the furnaces at the works, but as he could not always get access to those, he built his own small furnace at home to continue his researches.

He worked closely with physicists to investigate the properties of metals and in 1882 he made the important breakthrough when he made manganese steel. The alloy was 86.5 per cent iron, 1 per cent carbon and 12.5 per cent manganese. This was not an obvious choice of components. Manganese looks superficially like iron, but is brittle, rather softer and non-magnetic. It did not make the steel harder, but it made it extremely tough and resistant to cuts and abrasions. He went on to produce silicon steel, which turned out to have properties that were invaluable in another newly developing industry, electrical generation.

When his father died he took over the business, which was renamed Hadfield's Steel Foundry. There he continued to develop and improve his manganese and steel alloys, with the help of a team of expert metallurgists. The key to success was his own unflagging work rate, regularly putting in sixteen hour days. The business prospered, especially in the development of armaments. His steel was used to make armour-piercing shells and cast steel armour plating.

It found more peaceful uses in all kinds of ways, where the criterion was to withstand hard wear, such as railway crossings, where the manganese steel was found to last up to six times as long as conventional steel rails. The business that he had started in 1888, with a capital of less than £100,000 and almost 500 workers, had grown by 1918 to one that had a capital of nearly £2 million and was employing 13,000 workers. He was generally recognised as a good employer, one of the first to pioneer the eight-hour working day.

William Barrett is probably best known to the general public for his work on psychic research, but he was also an eminent physicist who spent a great deal of time investigating the magnetic properties of metals and metal alloys. Working with Hadfield, he discovered that alloys of silicon and steel had very useful properties. They were permeable to magnetism and had very low hysteresis. The latter is the property that defines the lag that occurs between the application and removal of a magnetic force. The magnetised material does not immediately return to its original state, resulting in a loss of energy, the hysteresis loss. Many components used in electricity generators are subject to variable magnetic fields, so the lower the hysteresis, the more efficiently they act. The new alloy proved to be ideal for electric components and was soon on the market as 'Stalloy'.

Hadfield was not the only one working on developing new alloys. The next inventor to make a major discovery of great commercial importance was also born in Sheffield in 1871. But Harry Brearley's background could scarcely have been more different than that of the rather austere, patrician Hadfield. His father was an ordinary steel worker and, according to his son in his book *Knotted String* (1941), 'an expert ale-supper'. His mother took in washing to help keep the family. He had an elementary education in the local board school, and he left at the age of 12 to work with his father at the Bessemer furnaces at Firth's Works in the city. There must have been something special about the boy, because the company's chemist encouraged him to study metallurgy, and by the time he was 20 he was able to move from the factory floor to take a job as a lab assistant. He left Firth's briefly, but when the company merged with the powerful John Brown enterprise he returned, and by 1907 he had been appointed director of the Firth-Brown Research Laboratories.

During a series of experiments in 1912 with different combinations of metals, basically intended for improving rifle barrels, he created a new alloy based on low carbon steel and 12 per cent chromium. Any new material could, by this time, be investigated under a microscope, and in order to provide a sample it would normally be etched with acid to provide a clean, smooth surface. He found that the acids he normally used had no effect. It appeared to him that an acid-resistant metal would have real commercial possibilities, but the company were not interested.

Brearley realised there was one obvious application. Conventional cutlery was easily blunted by the action of food acids, so he approached a local cutlery firm, George Ibberson, who was the first to use the new material. It was known as 'rustless' steel at first, but soon acquired the name by which it is still known today – 'stainless' steel. There has been some dispute over who first made this particular alloy, with claims being made for metallurgists in both America and Germany, but it does seem that he can certainly take credit for being the inventor of stainless steel cutlery.

The start of the European war in 1914 inevitably brought great changes to the steel industries of countries like Britain. As steel was essential for the armaments industry it was recognised that the industry would need some form of central control. An Iron & Steel Department was set up within the Ministry of Munitions, and was a success in that it actually managed to raise steel production to its highest ever level in spite of the extraordinary difficulties facing the industry.

In order to increase steel production, it was first necessary to make more pig iron and that, in turn required a regular supply of suitable ore. The problem that Britain faced was that much of the best quality ore had been imported from Sweden and Spain, and supplies were threatened by a new menace – submarine warfare. The German U-boats were having a devastating effect on merchant shipping. This not only depleted imports but also created an even higher demand for steel for building new ships to replace those that had been sunk. By 1916 it had become obvious that Britain would have to start producing more ore from old workings and develop new mines.

As in every industry at that time, there was a severe manpower shortage, with so many men away at the Western Front. When new ironstone mines were opened up, some 1,500 prisoners of war were set to work in them. It was not a satisfactory scheme. A few of the prisoners were Germans who had been living in Britain at the start of the war and were seen as potentially dangerous aliens, but the majority were captured German soldiers. It was hardly expected that they would show much enthusiasm in helping to provide the raw material for the shells that would be landing on their former comrades!

Even if they had been willing, the ministry responsible showed the kind of rigidity to rules that seems to afflict a certain type of bureaucrat. They insisted that they should only have the basic ration allowance given to all prisoners. It was sufficient for men with nothing to do but while away their time in a camp, but hopelessly inadequate for those doing hard, manual labour. Attempts to tempt workers from other industries failed.

What could have been a disaster turned out to become a long-term benefit. With labour in such short supply, the industry had no alternative than to mechanise. Steam diggers were introduced to opencast iron ore mines, railway tracks were laid to connect the mines to the main transport routes and trucks commandeered from wherever trucks could be spared. This proved to be a nightmarish operation, trying to find the correct rolling stock for the right situation. It was more than a little disconcerting for the manager of a mine only able to load low-sided wagons to be presented with a rake of high-sided trucks. But the job was done.

A new mine that opened up in Oxfordshire, for example, turned out to have the haematite that was ideal for the furnaces of South Wales, a branch of the industry that had previously relied on Spanish imports. By far the most important development was in Cumbria, particularly the vast Hodbarrow Mine on the coast near Millom. The sea virtually lapped at its door and could only be kept at bay by means of a great artificial barrier created out of blocks of furnace slag. By 1918, 5,000 men were employed in the iron mines of Cumbria.

The steel industry also developed despite suffering manpower shortages. The argument that a skilled man might be better employed manufacturing the essential armaments, rather than firing a single rifle, fell on deaf ministerial ears. Nevertheless, between 1916 and 1918, twenty-two new blast furnaces and 166 steel furnaces, mostly open hearth, were built in Britain. And just as problems in the ore-mining industry had forced change, so too changes were needed in the steel industry as a whole. With raw materials in drastically short supply, the country could not afford to waste any metal, and that included the swarf, the fragments of metal removed during machining, using lathes and drills. Neither the Bessemer nor the open hearth furnace could deal with the fine fragments of swarf, but an alternative was available: the electric arc furnace, developed as early as 1887.

Sir William Siemens had first demonstrated that an electric arc struck between two electrodes generated a very high temperature. The white heat had been used to provide illumination in the arc lamp, but it could also be adapted for heating furnaces. It was a Frenchman, Paul Heroult, who first used it for making steel in 1900 and it was to prove ideal for melting swarf.

Modern blast furnaces at Scunthorpe.

The great advantage of using electricity as a heat source was that there was no fuel involved and therefore no possibility of contamination. Electricity could also be used in the induction furnace. Induction was already used in transformers, in which a current in one circuit induced a current in a secondary circuit. The metal to be treated formed the secondary circuit, and the heat generated in the process was sufficient to melt the metal. The world of steel manufacture was moving forward, and the twentieth century was to see many new ways of using the material.

The motor car industry that had originated with the pioneering efforts of Gottlieb Daimler and Karl Benz was developing rapidly. But while car performance was improving steadily, production techniques had not really kept pace. In the first decades, cars were still built on the spot, with components being brought to the car, a system that was changed forever with the introduction of the production line by Henry Ford. One aspect of the work, however, was ripe for modernisation. The coachwork still looked back to the days of horse-drawn vehicles, with metal panels being laboriously attached to wooden frames. The man who instituted change was Edwin G. Budd.

Budd was born in Smyrna, Delaware, in 1870. His father was a Justice of the Peace, but Budd's interests were all centred on machinery. Instead of pursuing any sort of academic career, he became an apprentice machinist. His career as an inventor and innovator started, not with his own work, but with that of a friend, Thomas Corsicarden, who had devised a method of pressing steel sheets into different shapes. His invention was taken up by the American Pulley Co. and Budd soon joined that firm.

Apart from making pulleys, they also manufactured a range of castings in steel, including pedestals for another company, Hale & Kilburn, who specialised in making metal seating, especially for railway stations. The latter were interested in using pressed steel, and invited Budd to join the company to develop the project. His real breakthrough in manufacture came when he devised a method of joining the pressed steel panels to each other and to the basic frame by means of oxyacetylene welding.

This was exactly the technology that was needed to modernise car manufacture. In 1909, Emil Nelson of the Hupp Motor Co. invited Budd to join them and help in the production of an all-steel car, without the usual wooden framework. The scheme was a success, and Budd went on to perfect the technique. The availability of wide sheets of metal made it possible to manufacture the entire side of a vehicle in a single pressing, and combined with spot welding, the modern industry was born – and with it an inevitable increase, yet again, in the demand for steel.

The use of welding to join metal sheets together or to attach them to frames, using either the oxyacetylene burner or, latterly, the electric arc welder, was a process that had obvious attraction for shipbuilders. Rivets added an enormous amount of weight to a ship, and the rivet heads were, in effect, thousands or even millions of little bumps on the hull, affecting the flow of water past the vessel. But, British shipbuilders who had, in the nineteenth century, led the world in innovation had become more cautious.

As in many industries, success had led to a certain complacency. The general feeling was that the country made the finest ships in the world, made more of them than anywhere else and there was no need to change well-proven methods for novel techniques. In America, things were very different. New technologies were developed rapidly, especially during the Second World War. A crash building programme was begun, using prefabricated units that were welded together. It was fast and efficient and they turned out standard units. The T2 tanker was the shipping equivalent of the Model T Ford, a basic vessel that did the job with no frills and embellishments. In 1942 and 1943, 27 million tons of freighters and tankers were launched.

Welding was a success, but only if carried out with care. The application of high heat followed by rapid cooling can yield to distortion and weak joints. There were a number of cases of accidents reported where the fault appeared to lie with bad welding techniques. Such cases only added to the prejudices of the old school of riveters.

The problems in America were mainly due to bad practices, in some cases criminally bad. Men on piecework had a habit of filling in gaps with used electrodes from their arc welders and adding a cosmetic weld over the top. When the work was done, everything looked quite satisfactory to inspectors and the fault only appeared when the seam opened up, sometimes with catastrophic results. This was a fault that could be easily remedied by a proper inspection procedure.

The other fault, known as 'brittle fracture', was quite different. A plate would suddenly and inexplicably shatter. The most dramatic event occurred in Boston, where the tanker *Ponagansett* was being repaired. Men were carrying out a perfectly routine job, welding a small clip to the deck, when without warning the entire ship split in two as bow and stern dropped away. Eventually metallurgists were able to identify the stress patterns that had caused the effect and had started with minute cracks that spread right through a plate. New design regulations were set in place to solve the problem.

In the second half of the twentieth century, welding became the norm, even in Britain. The workforce were spared the cacophony of the riveting hammers, but had new problems to face. They were now working with dangerously high temperatures, and there was still no real concern for safety. Men would still come to work in their normal clothes with very little extra protection; the working conditions may have been quieter, but they were certainly no more pleasant.

Mr G. McLellan, who worked on the Clyde, told his story to the Oral History Project of the McLean Museum in Greenock:

> Aye it was the fumes off the welding, I don't know if you've ever seen a welder working it's just a blue smoke comes off it. But it's alright if you're in a big place. But if you're in a wee confined compartment it just gathers and gathers and you just can't see from one side to the other … the fumes in the confined spaces it was terrible. You got nothing at all and your eyes, they just watered. Best thing to do was to drink a pint of milk, but if you wanted a pint

> of milk you had to go and buy it yourself. And if you were off on the sick you lost the day off your work but you never got paid for it.

America was also the home of another important invention in the steel industry. Canned food was big business, and the old style of rolling mill could not keep up with the demand for sheet steel. Americans took the obvious first step of mechanising the process and then took it a stage further with the introduction of the continuous wide 'strip mill'. The name really tells the whole story. The metal plate was produced as a continuous strip, many feet wide. It could either be cut into appropriate pieces or more usually coiled up into a giant roll for transport. It was first introduced in the 1920s, but did not reach Britain until 1958 when a wide strip mill was added to the Ebbw Vale Steelworks.

The changes in the production of pig iron in the years following the Second World War were all designed to increase efficiency rather than make any fundamental changes. The main improvements came in making better use of the ore, a process for which the industry introduced an ugly new word – 'beneficiating'. In processing, a lot of loose material and dust is produced that cannot be fed directly into the furnace because it would clog everything up, but it can be used if it is sintered, that is, formed into lumps. As with other processes, much of the work was mechanised and, in later years, computer controlled.

The steel industry saw the introduction, not just of improved techniques, but fundamentally new ones. Post-war Europe needed steel in large quantities to help in the process of regeneration, but not every country had immediate access to the necessary raw materials. Austria had a severe shortage of scrap metal, but they devised an improved method of converting iron from the blast furnaces into steel. It was not unlike the Bessemer process, except that now pure oxygen was blown across the top of the molten metal in a high-speed jet.

It was named, after its innovators, the 'Linzer Düsenverfahren' (LD) process, a name that proved far too cumbersome for most users, and which soon became simply the LD process. One of the great advantages of the new process was its speed – a whole charge of around 300 tons could be processed in little more than half an hour. In later years, the idea was refined, by allowing molten metal direct from the furnace to fall through a ring of oxygen jets. Unlike the LD process, there is no need to stop and recharge the vessel, and the work can continue for as long as the furnace produces the molten metal.

The continuous system obviously had great advantages, but also created problems. It had always been necessary to test the quality of steel during production, but now things were moving so fast that by the time an old-fashioned analysis had taken place it had already become irrelevant. Fortunately, however, scientific methods had advanced and spectroscopy offered an analytical process that kept pace with production. Computers have enabled all aspects of the industry to be improved: rolling mills, for example, have been able to work to ever finer tolerances, and even the most old-fashioned mills have made use of the new technology.

The Forge de Syam in France is old-fashioned in the sense that it still hand rolls the steel, passing the hot metal through the rolls from one worker to the next. It is both a museum and a working factory. It might seem that there would be no market for this old technology, but they have brought the computer age to the works. They can roll to any desired shape to very great accuracy, because when the client sends in the specifications they are translated into an image on the screen that can then be used to design and produce rollers that are specifically shaped for that particular job. This flexibility means that they can take on work that is too small for the big companies to bother with. When I visited, they were rolling shaped metal strips for the floors of Otis lifts.

The whole movement in recent years had been towards producing steels of very high quality under carefully controlled conditions, and in producing new alloys. A problem with the older methods had been the presence of minute quantities of gas within the metal that would inevitably form weak spots. Originally, this was solved by de-gassing, sealing molten metal in a vacuum chamber. The unwanted gases would be drawn out of the metal and could be removed. More sophisticated methods used a second furnace to re-melt the steel under a vacuum as part of a continuous process.

Improvements have now made it possible to produce steel of the very highest quality in a controlled environment. It is not just the technology that has changed in modern times: the whole centre of the industry has shifted. In the early days Britain, the country that had the first Industrial Revolution, had led the world in innovation. From the late nineteenth century onwards, other countries had taken the lead. The British industry is no longer an important player on the world stage. Official statistics, issued by the World Steel Association, show the productivity of the sixty-nine different countries that, between them, make 99 per cent of the world's stage. Britain now has less than 1 per cent of total steel output, and is not even the leading European manufacturer. Even America is no longer top of the list, with just 7.3 per cent.

The new production centre is Asia – Japan, India and South Korea are catching up with the United States, but China has roared ahead. At the beginning of 2014, the country was making just slightly over half of the world's steel.

It has been a remarkable story of seemingly endless change ever since Abraham Darby made that vital technological breakthrough. Returning to Coalbrookdale, one finds that in spite of all the advances, fundamentals remain the same. Next door to the museum that houses Darby's original furnace, from which the parts for the iron bridge were cast, is a modern foundry. Appropriately, as the neighbour of the site that went to work producing cooking pots, the modern foundry is also turning out a kitchen product, a device that had its origins in Sweden.

Dr Gustav Dalen was a Swedish physicist who won the Nobel Prize for his work in developing an acetylene lamp for lighthouses. Then he lost his sight in an accident and was forced to stay home. For the first time he realised how much time his wife had to spend lugging buckets of fuel into the house to feed the cooker and heating stoves. He decided to do something about it. He developed the cast-iron stove, which was well insulated and whose temperature could be controlled by a thermostat. He was, at the time, managing director of a company called Aktiebolaget Gasaccumulator, but decided the name was too awkward for marketing, so settled for 'AGA'. It was a huge success, and was soon being manufactured in Britain, where it has remained so popular that it has become identified with a *certain* type of family, and has even given rise to its own fiction genre – the 'AGA saga'. Parts for this cooker are still being cast at the Coalbrookdale Foundry.

There are obvious differences between the old Darby Works and the more modern neighbour, but it is the similarities that are so striking. This is still a furnace producing molten iron that is fed into moulds to produce castings. The first obvious difference can be seen as soon as you enter the site. Outside in the yard is a huge heap of everything from old gearboxes to cog wheels, and even a few broken old AGAs. It is the scrap metal that now forms the charge instead of iron ore.

So, instead of a blast furnace, there is a cupola furnace. The scrap is brought to the furnace and carried to the top by mechanical hoist, where it joins the coke, still used as a fuel, and a limestone flux. Small measured amounts of silica and manganese will also be added. The molten metal pours out in a flurry of sparks to fill an oversized bucket, and every so often a fork lift truck arrives to carry it away to the production line. Inside the furnace, the temperature is 2,000°C, but by the time it is used for casting it has cooled to 1,400°C. The furnace men of old would probably have had little more than wet sacking as a protection against the intense heat. Nowadays they have fireproof clothing and a visored helmet.

In Darby's day, the molten metal was poured into moulds set in the casting floor, but here the moulds are brought to the metal on a conveyor. As each mould appears below the reservoir, the belt stops briefly, a measured amount of molten metal is dropped in and the mould continues on its journey, with a few flames flickering over the surface which gradually die away as it moves on. Darby, one feels, would have approved of this far more efficient system.

The changes are real, but in its essentials this is still the same, almost magical-seeming transformation that dates back over 2,000 years, when man first heated coloured rocks and got a metal that would transform the world. The New Iron Age, which began with the humble cauldron, has revolutionised transport on land and sea, built the machines that created an Industrial Revolution and transformed the cities of the world with ever-taller buildings. Our world is a world of iron and steel.

SELECT BIBLIOGRAPHY

Bessemer, H., *Sir Henry Bessemer FRS: An Autobiography* (1905).
Bird, A., *Paxton's Palace* (1976).
Brearley, H., *Harry Brearley, Stainless Pioneer, Autobiographical Notes* (1989).
Burton, A., *The Rise and Fall of British Shipbuilding* (2nd ed. 2013).
Burton, A., *Thomas Telford* (1999).
Carnegie, A., *Autobiography of Andrew Carnegie* (1920, republished 2006).
Chrimes, M., *Civil Engineering 1839–1889* (1991).
Corlett, E., *The Iron Ship* (1974).
Dickinson, H.W., *John Wilkinson, Ironmaster* (1914).
Fairbairn, W., *Iron, Its History, Properties and Processes of Manufacture* (1861).
Gale, W.K.V., *Iron and Steel* (1969).
Misa, Thomas J., *A Nation of Steel: The Making of Modern America* (1995).
Morgan, B., *Railways: Civil Engineering* (1971).
Nasaw, D., *Andrew Carnegie* (2006).
Raistrick, A., *Dynasty of Ironfounders* (1953).
Rees, D. Morgan, *Industrial Archaeology of Wales* (1978).
Rolt, L.T.C., *George and Robert Stephenson* (1960).
Rolt, L.T.C., *Isambard Kingdom Brunel* (1957).
Schubert, M.R., *History of the British Iron and Steel Industry* (1957).
Singer, C., et al (eds.), *A History of Technology* (7 volumes, 1954–1978).
Straker, E., *Wealden Iron* (1931).
Tweedale, G., *Giants of Sheffield Steel* (1986).
Ware, M.E., *The Making of the Motor Car* (1976).
Woods, M., and A. Warren, *Glass Houses* (1988).

INDEX